“十四五”职业教育部委级规划教材
教育部国家职业教育专业教学资源民族文化传承与创新子库
“中国丝绸技艺民族文化传承与创新”配套双语教材
江苏省高等职业院校高水平专业群“纺织品检验与贸易”配套教材

中国丝绸技艺概论

Introduction to Chinese Silk Techniques

杨建慧　朱　圳◎主　编
王丽丽　李沛赢◎副主编
王　伟◎译

中国纺织出版社有限公司

内 容 提 要 /Summary

本书介绍了悠久的丝绸历史与文化，详细介绍了蚕桑与制丝技艺、丝绸织造技艺以及丝绸印染技艺，总结了丝绸类非物质文化遗产的传承与创新，让读者全面了解丝绸传统技艺、产品以及相关丝绸文化，感受中国丝绸传统文化之美，为丝绸产品走出国门、走向世界搭建桥梁。

This book introduces the long history and culture of silk, details the sericulture and silk making, silk weaving and silk dyeing techniques, and summarizes the inheritance and innovation of silk-type intangible cultural heritage. It aims at assisting its readers in understanding the traditional Chinese silk techniques, products and culture, feeling the beauty of the excellent tradition culture of Chinese silk, and making the silk products known and available to the rest of the world.

图书在版编目（CIP）数据

中国丝绸技艺概论 =Introduction to Chinese Silk Techniques：汉文、英文 / 杨建慧，朱圳主编；王丽丽，李沛赢副主编；王伟译 . -- 北京 ：中国纺织出版社有限公司，2024.12

“十四五”职业教育部委级规划教材 教育部国家职业教育专业教学资源民族文化传承与创新子库“中国丝绸技艺民族文化传承与创新”配套双语教材 江苏省高等职业院校高水平专业群“纺织品检验与贸易”配套教材

ISBN 978-7-5229-1283-7

Ⅰ. ①中… Ⅱ. ①杨… ②朱… ③王… ④李… ⑤王… Ⅲ. ①丝绸－丝织工艺－中国－高等职业教育－教材－汉、英 Ⅳ. ①TS145. 3

中国国家版本馆 CIP 数据核字（2023）第 253556 号

责任编辑：由笑颖 孔会云　　责任校对：高 涵
责任印制：王艳丽

中国纺织出版社有限公司出版发行
地址：北京市朝阳区百子湾东里A407号楼 邮政编码：100124
销售电话：010—67004422 传真：010—87155801
http：//www.c-textilep.com
中国纺织出版社天猫旗舰店
官方微博 http：//weibo.com/2119887771
北京通天印刷有限责任公司印刷 各地新华书店经销
2024年12月第1版第1次印刷
开本：787×1092 1/16 印张：8.5
字数：173千字 定价：88.00元

前 言 Foreword

中华民族有许多伟大的发明创造，为全人类的文明进步做出了重大贡献，丝绸是最早和最伟大的发明之一。在古今中外的纺织品系列中，丝绸以其无穷魅力被喻为“纺织品皇后”，丝绸承载着许多中华文化的内容，蕴含着高贵典雅的审美品位，弥漫于中国的发展进程。随着“一带一路”倡议的实施，丝绸之路、丝绸文化、丝绸产品成为社会关注的焦点。丝绸不仅丰富了人们的服饰材料，美化了人们的生活，而且促进了中国和其他国家的交流。本书通过简单易懂的语言和精美的图片，介绍丝绸历史与文化、缫丝技艺、织造技艺、印染技艺、丝绸类非物质文化遗产这五个模块，让广大专业技术人员和丝绸文化爱好者了解独具中国特色的整个丝绸产业链，感受民族历史文化的魅力。

Among the great Chinese inventions that have made significant contributions to the progress of human civilization, silk is one of the earliest and greatest one. In a wide range of textiles of ancient and modern times, home and abroad, silk has been ranked at top of the textiles with its great charm. It is unique to Chinese culture, expressing noble and elegant sentiments throughout the entire development process of China. With the implementation of the Belt and Road Initiative, the Silk Road, silk culture and silk products have become the focus of attention in society. Silk not only enriches people' s clothing materials and beautifies people' s lives, but also facilitates the exchanges between China and other countries. With accessible language and pictures, this textbook introduces five modules: silk history and culture, reeling, weaving, printing and dyeing, and intangible cultural heritage of silk. Hopefully, the readers with professional and technical background and those interested in silk culture could acquaint themselves with the entire silk industry chain with unique Chinese characteristics and feel the charm of the national history and culture.

本书由苏州经贸职业技术学院杨建慧负责组织团队编写，中文部分第一章和第二章由杨建慧完成，第三章由王丽丽完成，第四章由李沛赢完成，第五章由朱圳完成，最后由杨建慧和朱圳统稿，英文部分由王伟完成。

This book was written by a team led by Yang Jianhui from Suzhou Institute of Trade and Commerce. The first and second chapters of the Chinese section were completed by Yang Jianhui, the third chapter was completed by Wang Lili, the fourth chapter was completed by Li Peiying, and the fifth chapter was completed by Zhu Zhen. Finally, Yang Jianhui and Zhu Zhen collaborated on the manuscript, while the English section was completed by Wang Wei.

因笔者水平有限，疏漏之处在所难免，恳请广大读者批评指正。

Due to the limited knowledge of the author, the shortcomings and mistakes of this book are inevitable. The readers are encouraged to make comments and suggestions for any revisions.

编者

2023年12月

目　录 Contents

第一章 丝绸的历史与文化 Silk History and Culture

丝绸，是我国古代最具中华民族特色的发明创造之一，承载着华夏文化的历史脉络，是中华文明的象征。中国是世界丝绸文明的发源地，有着非常深厚的历史渊源。华夏先民在6000年前开始种桑养蚕、缫丝织绸，创造了早期的丝绸文明，孕育了丰富的丝绸文化。

Silk, a symbol of Chinese civilization as well as one of the greatest inventions with national characteristics in ancient China, from which the world' s silk civilization originated, has rich historical veins of the development of Chinese culture. Approximately 6,000 years ago, the Chinese ancestors planted mulberry, raised silkworms, and reeled silk to weave, shaping an early silk civilization and a rich silk culture.

第一节 丝绸的历史/Silk History

蚕丝绸起源于中国，距今已有5000～6000年的历史，在史前时期就已经有蚕桑丝绸的存在，1926年，在山西省运城市夏县西阴村的仰韶文化遗址中，发掘出半枚被刀子切割过的蚕茧，这是能借以考证蚕丝起源于中国唯一的实物凭证。1958年，在浙江省湖州市吴兴区钱山漾新石器时代遗址中，出土了丝线、丝带等丝织品（图1–1），其中的丝织品是以捻合的家蚕长丝为经纬线交织而成的平纹织物，其表面细致光洁，丝缕清晰，是世界早期丝绸文明的重要证据。

Originating in China 5,000 or 6,000 years ago, silk already existed in the prehistoric period. In 1926, half of a silkworm cocoon cut by a knife was unearthed at a Yangshao Cultural Site in Xiyin Village, Xiaxian County, Yuncheng City, Shanxi Province. This was the only evidence that could verify the origin of Chinese silk at that time. In 1958, silk threads, ribbons and other silk fabrics were unearthed at a Neolithic site in Qianshanyang, Wuxing County, Huzhou City, Zhejiang Province (Figure 1–1). The silk fabrics are plain weave from the twinning filaments with

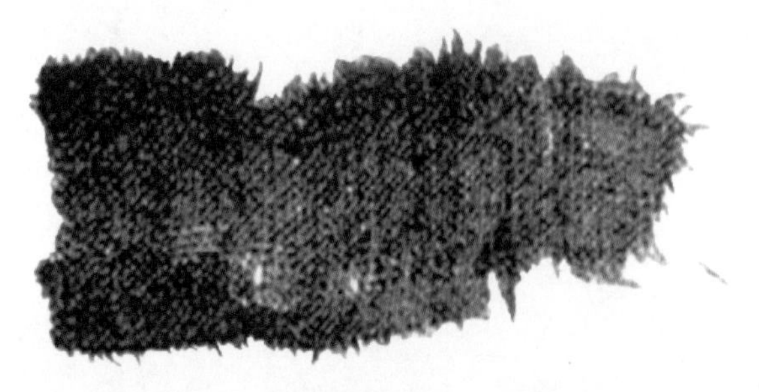

图1-1　钱山漾遗址出土的丝织品
Silk Fabrics Unearthed at Qianshanyang Relic Site

warp and weft knitted, which are important evidence of the early silk civilization in the world. Their surface is fine and clean, and the threads are clear.

除丝织品和纺织工具外，在新石器时代文化遗址中，还发现了许多蚕蛹形装饰物。1960年，山西省运城市芮城县西王村的仰韶文化遗址中，发现一个陶制的蚕蛹形装饰物。陶制蚕蛹长1.8厘米，由6个节体组成。1963年，江苏省苏州市吴江区梅堰良渚文化遗址出土的黑陶器上，绘有蚕纹图饰。到了西周时期，中国丝绸及其生产又是一番新的景象。1975年，陕西省宝鸡市茹家庄出土了两座西周墓葬，出土的文物中有玉蚕项链和丝织品。这些丝织品或附在铜器上的印痕里，或附在尸骨下的淤泥中，堆叠了三四层，可见数量之多，品种有绢、经锦和刺绣。

In addition to silk objects and textile tools, a few silkworm-shaped ornaments were found in the Neolithic cultural relic sites. In 1960, a pottery silkworm-shaped ornament with 1.8 cm in length and six nodes was discovered at the site of the Yangshao culture in Xiwang Village, Ruicheng County, Yuncheng City, Shanxi Province. Black pottery vessels unearthed at the Liangzhu Cultural Site in Meiyan, Wujiang County, Suzhou City, Jiangsu Province, in 1963, were painted with silkworm patterns. In the Western Zhou Dynasty, Chinese silk and its production was a new scene. In 1975, two tombs of the Western Zhou Dynasty were unearthed in Rujiazhuang, Baoji City, Shaanxi Province. Among the relics unearthed are jade silkworm necklaces and silk textiles. The silk textiles, such as silk, warp brocade and embroidery, are stacked in three or four layers in the imprint attached to the bronze utensils or in the mud attached to the bones of the corpses.

周朝重视农桑，再加上冶铁技术的发明和铁制农具的应用，社会生产有了迅猛的发展。这时，官营和民办丝织生产同时开展。西周时期，设有专门管理手工业的“工官”，丝绸生产从征收原料，到织、练、染，各种工序都设置“工官”管理：典妇工，掌管妇女织造；典丝工，负责征集蚕丝；染人，负责染丝染帛；设色工，又分画工、缋工、钟工、筐工、㡃工五个职位；另有负责征集植物染料的职位。每逢四时八节，王室和贵族们都要穿上丝织的礼服，各种仪仗旌旗都要用丝绸制作，并有指定的官员管理。周朝政府还制定了许多促进蚕桑生产的政策，规定将蚕丝作为赋税征收，因此老百姓不得不从事蚕桑丝织。

Mulberry planting was valued in the Zhou Dynasty, coupled with the invention of iron smelting technology and the application of iron farm tools, so that the silk production could develop rapidly. The official and private silk industries were run at the same time. There were then "industrial officials" who specialized in the management of handicrafts, responsible for various processes of silk production ranging from the collection of raw materials to weaving, scouring and dyeing, among whom are Dianfugong, the person in charge of women' s weaving; Diansigong, silk collection; Ranren, silk dyeing; Shesegong, sketching and embroidery, with five specializing positions: Huagong, Kuigong, Zhonggong, Kuanggong and Manggong. There were also positions in charge of collecting herbal dyes. To celebrate the switches of four seasons and eight solar terms of traditional Chinese calendar, members of the royal family and aristocracy wore silk dress, coupled with all kinds of ceremonial flags made of silk, all managed by specified officials. Many policies formulated by the government to promote sericulture production stipulated that silk should be taxed. Consequently, ordinary people had to engage themselves in silk weaving.

春秋战国时期，诸侯林立，列国争雄，各国都奖励蚕桑生产，将其作为富国利民的要策，因此各地丝绸业呈现出一派兴旺景象。从地域分布来看，丝绸生产遍布各诸侯国，中原地区胜于江南；从生产水平来看，养蚕方法讲究，缫丝质量很高，其纤维之细之匀可与近代产品相媲美；从丝绸品种来看，有十几种之多，如"齐纨""鲁缟""卫锦""荆绮""楚练"，都是当时独具地方特色的丝绸精品。

During the Spring and Autumn Period and the Warring States Period, all countries rewarded sericulture production, a preferential policy benefiting the country and its people, thus making the silk industry in various places prosperous. Geographically, it spread across an the vassal states, with the Central Plain region more popular than the regions south of the Yangtze River. In terms of production, the rearing of silkworms was particular and its fine fibers made by unreeling with warm water were comparable to modern products. As regards to silk variety, there were more than a dozen types of silk, such as Qiwan and Lugao in Shandong, Weijin in Henan, Jingqi in Hubei, and Chulian in Hunan, all of which were crafted with unique local characteristics.

从汉朝开始，由于丝绸之路西域道的开通，丝绸大量输出，成为西方各国特别是罗马上层社会的流行面料，由此奠定了中国丝织品的国际地位。为了发展蚕桑生产，汉朝专门设置了丝织机构，在京城长安设立"东织室"和"西织室"，在河南和山东分别设立"三服官"，在四川设立"锦官"，各拥有数千名织工，专门为皇宫、王室织造上等丝织品。与春秋战国时期相比，汉朝时丝绸的品类更加精细化，也开发出更多的品种。

Due to the opening of the western path of the Silk Road since the Han Dynasty, a large amount of Chinese silk had been exported, one that had become the popular fabric among western countries, especially the upper class of Rome, which had earned Chinese silk textiles international fame. In order to develop sericulture production, the Han government set up specialized silk weaving institutions, including the East Weaving Office and West Weaving Office in capital Chang' an, Sanfuguan (officials for the production of three types of seasonal silk clothing) in

Henan and Shandong, and Jinguan (officials for the production of Song brocade) in Sichuan. These specialized silk weaving institutions, each with thousands of craftsmen, wove fine silk for royalty. Compared with the Spring and Autumn Period and Warring States Period, the classification of silk was more refined in the Han Dynasty, with more varieties developed.

1959年，在新疆民丰县北部的一座东汉墓葬中发现大批织物，有三种是织山格文的平纹经锦。其中有一种用红、白、宝蓝、浅驼、浅橙五色丝线织成的“延年益寿大宜子孙”锦，据研究，这块织锦需要75片提花综才能织成，是当时制作最复杂的织物之一。另外两种是用红、白、紫红、淡蓝、油绿五色丝线织成的“万世如意”锦袍（图1–2），出土的各色绢大面积染色均匀、纯正，是东汉染品的代表作。

In 1959, a large number of fabrics were found in an Eastern Han burial in the north of Minfeng County, Xinjiang. Among the fabrics, there are three kinds of warp brocade with plain weave. A special brocade woven with Chinese characters “YAN NIAN YI SHOU DA YI ZI SUN” is made with colorful silk threads: red, white, sapphire blue, camel and light orange. According to research concerned, the brocade is made of 75 pieces of jacquard weave, one of the most complicated fabrics made at that time. The other two are brocade robes with “WAN SHI RU YI” woven from red, white, purplish red, light blue, and green silk threads (Figure 1–2). The unearthed silks with various colors are uniformly dyed in large area and looks pure,which are typical crafts of Eastern Han Dynasty dyeing products.

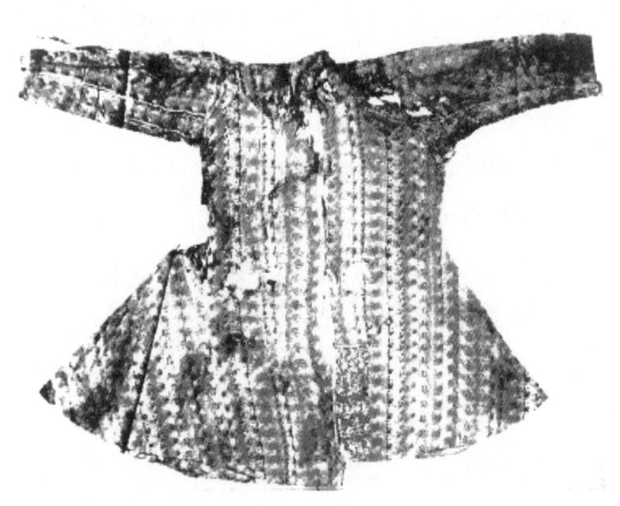

图1–2　东汉“万世如意”锦袍
“WAN SHI RU YI” Brocade Robe of the Eastern Han Dynasty

1972年发掘了长沙马王堆汉墓一号墓，出土的丝织品绚丽多彩，共有114件，大部分保存完好。这些丝织物中，衣着、鞋袜、手套等服饰约40件；杂用丝织物约20件；整幅或不成幅的丝织物约50件。马王堆汉墓还出有一种“起毛锦”的织物，可见汉朝织造发明了新工艺，而且工艺水平达到新的高度。同墓出土的一件素纱禅衣（图1–3），这件素纱禅衣为平纹交织，所用原料的线密度较细，质量仅49克，如果除去袖口和领口较重的边缘，质量仅25克左右，折叠后甚至可以放入火柴盒中。这代表了汉朝养蚕、缫丝、织造工艺的最高水平。

The No.1 Tomb of Mawangdui Han Dynasty mausoleum in Changsha City was excavated in 1972. There are 114 pieces of silk fabrics unearthed in total, most of which are well preserved. Among these silk fabrics are clothes, shoes, socks and gloves, about 40 pieces in total, together with about 20 pieces for trivial use, 50 with full or partial width. A kind of hairy brocade fabric is also found, suggesting that a new weaving process was invented in the Han Dynasty, and the craftsmanship reached a new stage. A piece of clothing with plain silk (Figure 1–3) unearthed has a plain-weave knitted structure, and the raw material has a small Linear density and weighs only 49 grams. If the edges of the cuffs and collars are removed, the weight is only about 25 grams, which can be placed in a matchbox after being folded. It represents the highest level of silkworm rearing, silk unreeling and weaving.

图1–3　马王堆出土的汉朝素纱禅衣
The Plain Silk Fabric of the Han Dynasty Unearthed in Mawangdui Mausoleum

在三国、两晋、南北朝时期，丝绸的生产受到战乱的摧残，但也受到一些有识之士的奖励和扶持。在当时税收制度的影响下，丝绸生产增长，丝织技术进步，再加上当时商业发达，商品流通量大，总的说来丝织业有一定的发展。1960年，出土了这个时期生产的富有东方色彩的绞缬和夹缬。绞缬是将白色丝绸按图案排列的要求用线扎结成块，然后染色，因为扎结处染料溶液不能渗入，染完之后扎结处就形成白色花纹。夹缬是用两块镂空的木板，将丝绸夹在当中，然后染色，利用木板夹紧处染料溶液渗透不良的原理，染出木板上镂刻的花纹。1966～1969年的发掘工作中，出土了西凉（400～421年）的绞缬和蜡缬。蜡缬是用蜂蜡在丝绸上手绘纹样，然后染色，点蜡处染液不能渗入，即形成白色花纹。由于蜡的龟裂现象，白色花纹上还有粗细不等的冰纹，呈现出淳朴而庄重的效果。目前这种产品在贵州的一些少数民族地区仍很流行，在东南亚和非洲也很受欢迎。

In the Three Kingdoms Period, the Western and Eastern Jin Dynasties, the Southern and Northern Dynasties, silk production ravaged by the war, was rewarded and supported by some elites. Under the influence of the tax system at that time, silk production increased and advances in technology emerged. Coupled with the development of commerce and the large amount of commodity circulation, there was a certain degree of development in general. In 1960, silk objects produced in these period with tie-dye and board-dye skills were unearthed. The former refers to knotting the white silk with threads according to the patterns required and then dyeing, since the dye solution does not penetrate the knots, and white patterns are formed after dyeing. Board-dye technique differs greatly. If a piece of silk cloth is sandwiched between two hollow carved wooden board and then dyed, the silk is dyed with the pattern carved into the board, because the dye solution does not penetrate where the board is clamped. The silk remains made from tie-dye and wax-dye techniques in the Western Liang Dynasty (400-421) were unearthed during the archaeological excavations from 1966 to 1969. If a piece of silk cloth is hand-painted with beeswax, and then dyed, white patterns appear because of no penetration into the wax spot. Due to the cracking phenomenon of wax, there are ice patterns with varying thicknesses on the white patterns, simple and solemn. The wax-dye still gains popularity among people in ethnic minority areas of Guizhou and finds favor with those in Southeast Asia and Africa.

唐朝丝绸是中国丝绸的又一座高峰。唐朝的桑蚕养殖更加普遍，丝织品种、产量、质量都十分惊人。唐朝政府十分鼓励蚕桑丝织，曾推行过“均田制”和“租庸调制”两项重要制度，这两项制度都与蚕桑丝织有关。为了更好地管理丝绸生产，唐朝中央政府还专门设立了织染署。唐朝的中央织染署分工很细，组织也很庞大，共设立了二十五个“作”，其中织细之作即有十个：布、绢、纱、绫、罗、锦、绮、褐等；组绶之作有五个：组、绶、绦、绳、缨；细线之作有四个：紬、线、弦、网；练染之作有六个：青、绛、黄、白、皂、紫。

Silk in the Tang Dynasty is another higher stage of Chinese silk. Silkworm rearing was more popular in the Tang Dynasty, where there were an amazing number of silk products with wider range of types and of higher quality. The government highly encouraged sericulture, and once carried out two important policies: land equalization system and Zu-Yong-Diao system, both of which were related to sericulture. To better manage silk production, the central government also set up a special weaving and dyeing department. The department had a fine division of labor and a huge body of organizations, with twenty-five sections set up, among which there were ten fine weaving ones: cloth, spun silk, yarn, silk twill, half-cross leno, brocade, figured woven silk, coarse cloth etc.; five sections for official seal tapestry and silk strip ornaments: official seal tapestry, silk strip ornaments, silk ribbon, rope and tassel; four bourette-silk ones: noil silk yarn, thread, string and net; and six scouring and dyeing ones: green, crimson, yellow, white, black and purple.

唐朝的丝绸质量从唐朝初期就很高，从当时全国十个道每年向朝廷交纳的贡赋丝绸品种就可以看出来，如河南道交纳的方纹绫、双丝绫、镜花绫、仙文绫，河北道交纳的孔雀罗、春罗等，这些都是当时各地花式新颖、花色绮丽的高级丝织品。随着与中亚、西亚的文化交流更加频繁，唐朝工匠不断汲取西方纺织文化的营养，盛唐时期的丝织品没有了以往的神秘、细腻，也不再简约、古朴，丝绸的色彩更加艳丽豪华，图案以洋溢生活气息的花鸟居多，显示出一派春意融融、生机勃勃的景象。

The silk was of high quality from the beginning of the Tang Dynasty, which can be seen from the varieties of tribute silk paid to the court every year by ten regions in the country, such as Fangwen lea, Shuangsi silk twill, Jinghua silk twill and Xianwen silk twill by Henan, Kongque half-cross leno and Chunluo half-cross leno by Hebei, etc. These are all the high-quality silk fabrics with novel and beautiful designs in different places at that time. With the more frequent cultural exchanges with Central Asia and West Asia, the craftsmen of the Tang Dynasty acquainted themselves with the western textile culture. Exquisite and complex, the silk textiles of this period were well known to all, with patterns of flowers and birds showing a scene full of spring and vitality.

随着时代的发展，宋朝的社会和丝织生产展现了新的特点。首先，由于商品经济比先前发达，因此大城市店铺林立、农村也形成许多繁茂的小市镇。其次，民间的蚕丝生产和织绸生产开始分工。农家一般养蚕缫丝，而不自己织绸，只是把生产的蚕丝拿去出卖，由专门的“织帛之家”来织绸。从宋代开始，中国经济重心由北方逐渐转移到南方，丝绸业的中心也随之聚集于江南地区，江南丝绸业也逐步超过北方。

With the development of the times, the silk production in the Song Dynasty showed several characteristics. First, because the commodity economy was more developed, there were many shops in big cities and flourishing small towns in the countryside. Secondly, folk silk production and weaving production began their work on the basis of a farm-out mode. Farmers generally reared silkworm and reeled silk, but did not weave silk; instead, they only sold the silk thread. Since the Song Dynasty, China’ s economic focus had gradually shifted from the north to the south, and the center of the silk industry also gathered in the south of the Yangtze River, where the industry of the south surpassed the north.

江南的蚕桑业和丝织业已经远远超过黄河流域，当时的“两浙路”范围包括现浙江全省及苏南大部分地区，其所缴纳的丝织品占全国四分之一以上。宋代的罗织品生产水平更是达到新高，每年各地贡物中，罗织品达十万匹以上，其中江浙一带就占全国总数的六成以上。南宋建都杭州，大大刺激了太湖流域蚕桑和丝织业的生产，著名的宋锦就产生于这个时期的苏州。宋代丝绸中心南移奠定了我国长江下游丝绸业的中心地位，开拓了明清以至现代南方丝绸繁盛的历史新局面。

The sericulture industry and silk weaving industry in the south of the Yangtze River had far surpassed their counterparts of the Yellow River basin. At that time, the Liangzhe Road (an administrative region) covering the entire Zhejiang and most parts of the southern Jiangsu accounted for more than a quarter of the whole country. The production of half-cross leno textiles in the Song Dynasty reached higher level, with its output soaring to more than 100,000 pi (1 pi equals 13.2 meter or so) per year, among which Jiangsu and Zhejiang accounted for over 60%. Hangzhou, the capital of the Southern Song Dynasty, greatly stimulated the production of sericulture and silk weaving in the Great Lake basin. The famous Song brocade was produced in Suzhou during this period. The southbound movement of the silk center in the Song Dynasty established the central position of the silk industry in the lower reaches of the Yangtze River, a new stage of the prosperity of silk in the Ming and Qing Dynasties as well as the modern south.

明朝时期，丝绸业和蚕桑丝织生产的商品化日益发展，丝绸开始主要向南洋、日本、朝鲜等国家和地区输出，后来逐渐转变为主要向欧洲、葡萄牙、西班牙、荷兰等国家和地区出口。我国的蚕桑丝织，一方面，在社会需要下高度发展，达到历史上前所未有的社会化分工，尤其是在南方江浙一带，出现了空前的繁荣。另一方面，明朝中后期，落后的封建制度又重重地约束着生产力的发展。官府的重税和专卖政策阻碍着手工业和商业的自由发展。岁办、额办、买办、杂办等名目繁多的征收项目，使丝织机户不堪重负，还有各种织染局，以低价硬派机户织造，仅浙江就在杭州、绍兴、金华、台州、温州、宁波、嘉兴、湖州等多处设有这种机构。这样，不仅制约了我国丝绸业的发展，而且促使我国丝绸业出现一个新的特点，即生丝的生产和出口几乎超过了绢绸的生产和出口。外国需要中国雄厚的蚕丝业作为其织造的后盾，而中国的蚕丝生产在国外大量需求下又急速增长。因为生丝贸易比绢绸贸易的利润高至数倍，浙江和福建的一些商人不顾朝廷的禁令，冒险卖生丝，牟取厚利。

The commercialization of silk industry and sericulture production gradually developed in the Ming Dynasty, when silk was exported mainly to South Asia, Japan, Korea and others at first, and then gradually to Europe, Portugal, Spain, the Netherlands, etc. China' s sericulture and silk weaving was highly developed according to the social needs, achieving an unprecedented social division of labor in history, especially in the southern areas of Jiangsu and Zhejiang, where prosperity took place. However, the backward feudal system in the middle and late periods restricted the development of productivity. The government' s heavy tax and monopoly policies hindered the free development of handicraft industries and commerce. There were numerous expropriation items such as yearly tribute, quota tribute, comprador, and additional tribute, which overwhelmed the silk weaver households. There were also various weaving and dyeing offices that sent machines to households to weave, with the silk products sold at low prices. Zhejiang alone had such institutions in Hangzhou, Shaoxing, Jinhua,

Taizhou, Wenzhou, Ningbo, Jiaxing, Huzhou and others. As a result, this plagued the silk industry in China in that the production and export of raw silk almost exceeded that of silk textiles at that time, a new but worried characteristic of the industry. China' s strong silk industry was needed by other countries to support their silk weaving industry, thus booming China' s raw silk production under the growing demands abroad. Since the profit of the raw silk trade was several times higher than that of the silk trade, some merchants in Zhejiang and Fujian who sold the raw silk by sea took risks to get huge profits, despite the court' s ban.

清朝前期的蚕丝生产，基本上属于家庭副业的范畴，农民出售的商品是生丝。由于祖辈世代养蚕，积累了丰富的经验，特别是江浙地区，勤于培桑，精于养蚕，长于缫丝，丝的质量非常好，收益甚好，超过了种粮。遇到丝价高涨时，生丝生产加速膨胀，甚至发生了良田种桑，畸形发展的情况。另外，清政府实施严格的海禁，禁止下海经商。康熙初年，海禁更严，将浙江、福建、广东诸省沿海居民内迁，筑界墙，禁止外国船只进入，准许少数外国船只停泊贸易。由于外商对中国丝绸的需求，清统治阶级对外洋珍物奇玩的爱好，以及考虑到关税是项巨大的收入，1683年开海禁，但对下海商船的大小进行了限制。从清初到鸦片战争时期，海禁多次恢复，特别是对丝货的输出，禁限更严，严禁民间丝货贸易的目的可能是官府能垄断其利，并防止生丝涨价、好丝输出，以免影响国内上等缎匹的生产等。

The production of silk in the early Qing Dynasty was not the households' primary livelihood, and what the peasants sold was raw silk. Since the silkworms were reared from generation to generation, the rich experience had been accumulated, especially in Jiangsu and Zhejiang, where the mulberry silkworms were well reared, and the quality of silk was excellent, with more profits than those of growing grain. When the price of silk rose, the production of raw silk accelerated, and even the planting of mulberry in fertile fields occurred. In addition, the Qing government implemented a strict sea ban, forbidding business. In the early years of Kangxi, the ban was even stricter, moving coastal residents in Zhejiang, Fujian and Guangdong inward, building boundary walls, banning foreign ships to enter, and allowing only a small number of foreign ships to berth for trade. It was imposed in 1683 because of the foreign demand for Chinese silk, the Qing ruling class' s hobby for foreign treasures and fun, and a huge income of tariff, with a limit still kept on the size of merchant ships. From the beginning of the Qing Dynasty to the Opium War, the ban was resumed many times. In particular, the export of silk goods was more stringent. The purpose of strictly prohibiting the trade of folk silk goods might be that the government could monopolize the benefits and prevent the price increase of raw silk and the export of good silk, so as not to affect the production of domestic high-quality satin pieces.

第二次世界大战期间，受世界经济危机和战争的影响，我国蚕丝业受到严重打击，江、浙的缫丝厂大部分被毁。此外，日本采用蚕丝业统制政策，对中国蚕丝业实行垄断、独占，以实现“包中国蚕丝业于日本蚕丝业的势力圈内”的战略。

During the World War Ⅱ, affected by the world economic crisis and war, the Chinese silk industry was hard hit, and most of the reeling silk factories in Jiangsu and Zhejiang were destroyed. Besides, Japan adopted the policy to monopolize the Chinese silkworm industry and take control of it in order to realize its strategy of covering the Chinese silkworm industry within the Japanese silkworm industry.

中华人民共和国成立以来，我国蚕桑丝织业得到了极大的恢复和发展，已踏上了振兴之路。江南地区仍然是我国首屈一指的蚕桑基地和丝织中心。蚕桑和丝绸生产多年居全国前列，不仅量多，而且质量好，工人和农民的经验丰富，技术熟练。四川的蚕丝业发展迅猛，产茧量和产丝量跃居全国前列。珠江三角洲蚕桑丝织业一片兴旺景象，桑基鱼塘，鳞次栉比，吸引着一批又一批海外游客考察观赏。

Since the founding of the People' s Republic of China, the domestic silk industry has greatly recovered and developed, and has embarked on the road of revitalization. The silk industry in the south of the Yangtze River has still been the leading sericulture base and silk weaving center in China. Both sericulture and silk production has been in the forefront of the country for many years, not only in large quantity but also in good quality, with workers and farmers experienced and skilled. The silk industry in Sichuan has developed rapidly, with its cocoon and raw silk production rising to the forefront place. The sericulture industry in the Pearl River Delta is booming, where a great deal of mulberry trees grow along the mulberry-based fishponds, a thriving picture that attracts overseas visitors to come.

丝绸外贸空前繁荣，丝绸内销兴旺发达，为国家经济建设积累着资金，并日益满足着人民的物质需求和精神需求。目前我国的丝绸远销世界的150多个国家和地区，在世界的各个角落都可以看到中国丝绸或用中国生丝织制、印染的绸缎。

Foreign trade in silk has flourished unprecedentedly, and its domestic sales thrived, accumulating funds for the country' s economic construction and increasingly meeting the people' s material and spiritual needs. China currently sells silk to more than 150 countries and regions on all continents of the world. It is no exaggeration to say that Chinese silk or satin woven and dyed from raw Chinese silk can be seen in every corner of the world.

第二节　丝绸之路/Silk Road

丝绸以柔顺的质地、绚丽的色彩、丰富的纹样，作为中国上层社会的奢侈品，象征着身份和地位，同时也为古代中国的对外交流做出了突出贡献。从汉朝开始，丝绸随商人、使者、僧侣、旅行家的足迹，甚至伴随征战者的铁马金戈，传到了中亚、西亚，传到了阿拉伯地区乃至欧洲，成为中国和西方国家进行大规模经济、文化交流的纽带，架起了中西交通的桥梁，并且被赋予了新的内涵。这座中西交通的“桥梁”以一个响亮的专有名词而著称于世，即丝绸之路。丝绸之路，广义上又分为陆上丝绸之路、草原丝绸之路和海上丝绸之路。

With its soft texture, gorgeous colors and rich patterns, silk, a luxury in the upper classes of society in ancient China, symbolized the identity and status, and also made outstanding contributions to ancient China' s foreign exchanges. From the Han Dynasty, along with the footsteps of businessmen, envoys, monks, travelers, and even those accompanying the conquerors in wars, silk spread to Central and West Asia, Arab region and even Europe. It became a link for large-scale economic and cultural exchanges between China and Western countries, a bridge between China and the West with new connotations. The “bridge” between China and the West is well known for its resounding proper noun, the Silk Road, which is broadly divided into the Land Silk Road, the Prairie Silk Road and the Maritime Silk Road.

一、陆上丝绸之路 /Overland Silk Road

提起丝绸之路，人们自然会联想到在古代陕甘高原通向西域的崎岖道路上，一队队由骆驼组成的商队，驮着油漆麻布或皮革制成的行囊，沿着张骞出使西域新开拓的路线向西行进的盛况。

When it comes to the Silk Road, one will naturally think of the grand occasion of a caravan of camels carrying bags made of painted linen or leather along the rugged road leading to the west from the ancient Shanxi-Gansu Plateau, following the newly opened route of Zhang Qian' s mission to the Western regions.

丝绸之路东起西汉的国都长安（今西安），途经甘肃的武威，穿过狭窄的河西走廊，到达当时的中西交通枢纽——敦煌，再往西就是塔克拉玛干大沙漠，因此，在这里需绕过大沙漠西行，路分为南北两条。南路西出阳关，经鄯善、于阗、莎车，再往北到达疏勒，又往西到达大宛。北路是西出玉门关，经车师前国、龟兹到达大宛。在大宛，南北两路会合继续往西到达安息、条支和大秦，“丝路”全长7000多千米。出兰州一直向西，漫长的丝绸之路大都是在戈壁沙漠中穿行，环境气候恶劣，人烟稀少，落后的交通工具和道路条件决定了丝路贸易的中转性，使沿线城郭都变成了大大小小的商品集散地，它们随丝路通达而兴，随丝路阻塞而衰，许多深埋于黄沙之下的古城，凝固了一段段繁华的历史。

The Silk Road starts from Chang' an (Xi' an), the capital of the Western Han Dynasty, through Wuwei, Gansu Province, the narrow Hexi Corridor, to the then transportation hub—Dunhuang. Further to the west is the Taklimakan Desert. Therefore, the road is divided into two routes from north to south, and it goes west around the desert. The south road goes out of Yangguan in the west, passes through Bushan, Yutian and Shache, and then reaches Shule in the north, and Dawan in the west. The north road is the west of Yumen Pass, through the ancient states of Cheshiqian, Qiuci to Dayuan. At Dayuan, the north and south roads meet and continue westward to Anxi, Tiaozhi, and Daqin. The Silk Road stretches more than 7,000 kilometers long. Out of Lanzhou all the way to the west, the long Silk Road mostly runs through the Gobi Desert. The harsh environment, sparse population, backward means of transport and road conditions determined the transit role of the Silk Road trade. As a result, the cities along the route became large and small commodity distribution centers, rising with the access and declining with the blocking of the Silk Road. Many ancient cities buried deep under the yellow sand reveal their prosperity over a long period time in history.

丝路贸易的繁荣和鼎盛主要在汉朝和唐朝，当时的长安、洛阳，河西走廊的敦煌、酒泉、张掖、武威，以及塔里木盆地周围地区于阗、疏勒、轮台、鄯善、楼兰、吐鲁番等，都是丝绸之路上有名的国际交易市场。

The Silk Road trade mainly boomed during the Han and Tang Dynasties. Among famous international trading markets along the route are Chang' an and Luoyang; Dunhuang, Jiuquan, Zhangye and Wuwei of Hexi Corridor; and the surrounding areas of the Tarim Basin, such as Yutian, Shule, Luntai, Shanshan, Loulan and Turpan.

丝绸之路出敦煌后就进入新疆，分为南、中、北三条路线。南线和中线各沿塔里木盆地的南、北边缘前进，汇合于喀什。南线自敦煌经罗布泊到楼兰，一路经过若羌、且末、民丰、于田、和田、莎车到喀什。中线自敦煌经艾丁湖到吐鲁番，过焉耆、库车、拜城、阿克苏再到喀什。从喀什往西越过葱岭即到大月氏和安息。北线则经哈密、吉木萨尔、乌鲁木齐、昌吉、霍城、阿拉木图到碎叶城，直到东罗马帝国首都君士坦丁堡。

Away from Dunhuang, the Silk Road enters Xinjiang and is divided into three routes: south, central and north. The south and middle routes advance along the south and north edges of the Tarim Basin and join in Kashi. The south route runs from Dunhuang to Loulan via Luobupo and passes all the way through Ruoqiang, Qiemo, Minfeng, Yutian, Hotan and Shache to Kashi. The south route from Dunhuang passes through Lake Aydingkol to Turpan, via Yanqi, Kuche, Baicheng, Aksu to Kashi. Moving westward from Kashi across Congling (the Pamirs), one arrives at the area where the Yuezhi once lived and then at the ancient state Anxi. The north route passes through Hami, Jimser, Urumqi, Changji, Huocheng and Almaty to Suyab until the eastern Roman imperial capital Constantinople.

张骞通西域之后，塔里木盆地南缘的于阗、楼兰、龟兹等地迅速成为丝绸贸易的中转站，西亚商人大多在这里获得从中原运来的丝绸，再转运到西亚各地。于阗、楼兰、吉木萨尔、哈密都是当时各国商人云集的西域丝都。古楼兰、吐鲁番出土的历代精美丝绸，于阗出土的各国钱币，如汉五铢钱、唐开元通宝、波斯银币、罗马金币等，都反映出这些古代名城昔日活跃的丝绸贸易。

After Zhang Qian' s mission to the west, Yutian, Loulan, Qiuci and other places on the southern edge of the Tarim Basin quickly became transit stations for the silk trade. Most of the merchants in West Asia obtained the silk from the Central Plains and then transferred it to various parts of West Asia. In addition to Yutian and Loulan, Jimser and Hami were all the Silk Road cities where businessmen gathered at the time. The booming silk trade was well presented in these ancient cities, such as the exquisite silk of ancient Loulan and Turpan unearthed and the coins of various countries unearthed, such as Wuzhu coin in the Han Dynasty, Kaiyuan Tongbao (ancient coins with the square hole in center), Persian silver coin and Roman gold coin.

二、草原丝绸之路 /Steppe Silk Road

草原丝绸之路是指蒙古地带沟通欧亚大陆的商贸大通道，是丝绸之路的重要组成部分。它从中原地区向北越过大青山、燕山一带的长城沿线，向西北穿越蒙古高原，翻越天堑阿尔泰山，再经准格尔盆地到哈萨克丘陵，或直接从巴拉巴拉草原到黑海低地至匈牙利。早在公元前5世纪前后，中国的丝绸产品就通过草原丝绸之路传至欧洲。草原丝绸之路的形成，在很大程度上与自然生态环境有着密切的关系。在整个欧亚大陆的地理环境中，东西方之间的交流是极其困难的。在整个欧亚大陆中，只有北纬40 ~ 50度的中纬度地区，才有利于人类进行东西向交通，而该地区正是草原丝绸之路的所在地。这里是游牧文化与农耕文化交汇的核心地区，因此，草原丝绸之路不仅沟通东西方向，而且沟通南北方向，在经济与文化交流中发挥着重要作用。

The Steppe Silk Road refers to the major trade channel between Mongolia and Eurasia, an important part of the Silk Road. This steppe route crosses the Great Wall along Daqing Mountain and Yanshan Mountain from the Central Plains to the north, the Mongolian Plateau to the northwest, climbs over the Altai Mountains, and then passes through the Junggar Basin to the Kazakh hills, or directly from the Barabara Prairie, the Black Sea Lowlands to Hungary. As early as around the 5th century BC, China' s silk products were exported to Europe through this route, one that was largely related to the natural ecological environment. In the whole Eurasian continent, the communication between East and West is difficult. Throughout Eurasia, only the mid-latitudes between 40 and 50 degrees north latitude is conducive to human east-west traffic, an area where the Steppe route is located and where nomadic culture and farming culture meet. Therefore, the route not only communicates between the East and the West, but also between the North and the South, whose role in economic and cultural exchange is very important.

三、海上丝绸之路 /Maritime Silk Road

海上丝绸之路，顾名思义就是运送丝绸的海上通道。它的得名是由于当时我国从海路出口的商品和陆上丝绸之路一样，相当一部分是丝绸。海上丝绸之路是当时海上航线的泛称，不是仅指某一条具体的航线。海上丝绸之路的起点在我国东南沿海，终点在非洲东北部埃及沿海港口。海上丝绸之路在历史上是一条重要的商业运输线，即使在今天也仍然是东西贸易的重要通道。海上丝绸之路的历史可以追溯到千百年前。

The Maritime Silk Road, as its name implies, it is the sea route that transports silk products. It bore the name because a significant portion of the exports from the sea route was also silk at the time, as was the case with the land routes. It was commonly known at that time as a general term, not just a specific sea route, whose starting point lies along the southeast coast of China, and which ends at the port of Egypt in northeastern Africa. The Maritime Silk Road has historically been a more important commercial transport route than the land routes. Even today, it is still an important channel for trade between the East and the West. The history of the Maritime Silk Road can be traced back thousands of years.

在张骞通往西域和陆上丝路形成之前，就已有了海上丝绸之路，而且有从东海至朝鲜和日本，从南海至地中海东非两条起航线。

It existed before the Overland Silk Road and Zhang Qian' s mission. There were also two routes from the East China Sea to North Korea and Japan, and from the South China Sea to the Mediterranean East Africa.

周武王灭纣，建立周王朝时，周武王封箕子出使朝鲜，从山东半岛的渤海湾出发，乘船前往朝鲜，传授养蚕和织丝的技巧。这是史籍中较早的东海起航线。隋唐时期，中日航线繁忙，丝绸贸易十分兴盛。因此，日本也保留了较多的唐代丝织品，一些精品甚至在中国大陆也难以见到了，如彩色印花锦缎、狩猎纹锦、莲花纹锦、狮子唐草秦乐纹锦等。

Having fought against Emperor Zhou of the Shang Dynasty, and established the Zhou Dynasty, Emperor Wu of Zhou sent Jizi (senior official of Shang) to Korea, the latter departing from the Bohai Bay on the Shandong Peninsula to Korea, where he introduced laws and rules, and taught skills of silkworm rearing and silk weaving. This is an earlier route ever found in historical records from the East China Sea route. During the Sui and Tang Dynasties, the Chinese and Japanese routes were busy, and the silk trade was very prosperous. This is why there are more Tang silk textiles kept in Japan, some of which are not accessible to the public in mainland China today, for example, various brocades with delicate patterns woven, including colored printing, hunting, lotus, lion, flower, etc.

南海丝路初始于南方丝路，即从四川出发到云南，从保山以南，沿伊洛瓦底江到仰光，再西行至印度，由印度渡印度洋；或经入口进入中亚，或继续沿海到达罗马帝国。到了汉朝，汉武帝一边由北向西域拓展，另一边由南向东南亚拓展，招募海员从广东徐闻、广西合浦港出海，经过越南沿海岸线西行，到达印度境内、斯里兰卡。他们带去的主要有丝绸和黄金等。这些丝绸再通过印度转销到中亚、西亚和地中海各国，但这条路线主要是汉朝时出于政治目的开辟的，并没有成为一条繁忙的商业航线。

The South China Sea route started from the southern Silk Road from Sichuan to Yunnan, south of Baoshan, along the Irrawaddy River to Yangon and then westward to India, cross the Indian Ocean from India, or by entrance into Central Asia, or along the coast to the Roman Empire. In the Han Dynasty, King Wu expanded to the west in the north and Southeast Asia in the south, recruiting seafarers, who were sailing away from the Xuwen port in Guangdong and Hepu Port in Guangxi. They traveled westward along the coast of Vietnam to India and Sri Lanka. The main goods brought with them were silk and gold. The silk was then sold through India to Central Asia, West Asia, and the Mediterranean countries, but the route was largely created by the Han government for political purposes and did not become a busy commercial one.

隋朝时朝，隋炀帝非常重视与东南亚各国的交流。在公元607年，他派遣使者常骏、王君政出使马来半岛上国土广大的赤土国，并带去了帛数百，缎五千，时服一批。其航线应是从广州出发，经过北部湾，到达越南最南端，再到马来半岛东南部，最后到达赤土国。

During the Sui Dynasty, King Yang attached great importance to the exchanges with Southeast Asian countries. In AD 607, he sent emissaries Chang Jun and Wang Junzheng to the vast land of the Chitu kingdom on the Malay Peninsula, who brought abroad hundreds of silks, 5,000 satin and a batch of fashion clothing. The route should be from Guangzhou, via the Beibu Gulf, to the southernmost tip of Vietnam, and then to the southeast of the Malay Peninsula, and finally to the Chitu.

自宋朝开始，中国的经济重心南移，江南和四川成为主要的丝绸产区，从东南沿海城市广州、泉州、杭州等地出发，海上航线日益发达，范围越来越广。从南洋到阿拉伯海，甚至远达非洲东海岸，海外贸易一片繁荣。印度、罗马、东南亚、东非等50多个国家的海船常常直接到中国南方各地采购丝绸。宋朝的造船技术已经相当发达，此时已经发明指南针并用于航海。20世纪80年代，在广东阳江海域发现了一艘宋朝的巨大沉船，命名为“南海一号”。这是一艘远洋货船，据初步探测，整船装载文物有5万～8万件，相当于一个省级博物馆藏品的总量，船体巨大，其运载能力是沙漠之舟的骆驼无法比拟的。这艘沉船的发现，展现了茫天沧海中帆影重重的海上丝绸之路昔日的繁华与沧桑。当时的贸易方式，一种为“朝贡贸易”，也就是外国商人以呈献当地物产为主，宋王朝以回赠丝绸等贵重物产作为答谢，回赠的价值往往远远超过贡物的价值。另一种为“市舶贸易”，即正式的交易，宋朝在广州、泉州、杭州、宁州、嘉兴、温州、江阴、山东诸城等地设立了市舶司，专门管理对外贸易，政府征收商业税，并鼓励中国商人进行丝绸、陶瓷和茶叶的海上贸易，从而促进珠江三角洲和福建地区丝绸生产的发展。

Since the beginning of the Song Dynasty, China' s economic center has been shifted southward, with the south of the Yangtze River and Sichuan province becoming the main silk-producing areas. Starting from southeastern coastal cities of Guangzhou, Quanzhou, Hangzhou and other places, the sea routes have become increasingly developed, which reaches farther and farther, from the Southeast Asia to the Arabian Sea and even the east coast of Africa. The overseas trade further flourished. Merchants traveled directly to southern China by ships to purchase silk from more than 50 countries and regions, including India, Rome, Southeast Asia, and East Africa. The ship building technology of the Song Dynasty has been quite developed, an ear when the compass was invented and used for navigation. In the 1980s, a huge shipwreck of the Song Dynasty was discovered in the sea area of Yangjiang in Guangdong, named "Nanhai No.1", which is an ocean-going freighter. According to preliminary exploration, the ship with a huge hull carries about 50,000 to 80,000 pieces of cultural relics, equivalent to the total amount of collection of a provincial museum, with a carrying capacity unmatched by the camels in the desert. The discovery of this shipwreck shows the prosperity and vicissitudes of the former maritime Silk Road. One way of trade at the time was the "Tribute Trade", in which foreign merchants mainly presented local goods, and the officials in the Song Dynasty reciprocated with precious goods such as silk, the value of which often far exceeded that of the tribute. The other is the "Market Trade", which is the official transaction. The Song Dynasty set up the Department of Shibo in Guangzhou, Quanzhou, Hangzhou, Ningzhou, Jiaxing, Wenzhou, Jiangyin and Zhucheng, specializing in the management of foreign trade. The government levied the commercial tax and encouraged businessmen to trade by sea in silk, ceramics and tea, thus promoting the development of silk production in the Pearl River Delta and Fujian.

元朝的陆上丝绸之路，由于元朝疆域辽阔，远至黑海、波斯湾地区，十分畅通，然而海上丝路更加繁荣。元朝时期，在泉州、宁波、上海、澉浦、温州、广州、杭州设立了七处市舶司，管理对外贸易。这些城市的港口船只熙熙攘攘，尤其是繁荣的泉州港，给许多外国游客留下了深刻的印象。元朝外贸实行的是“官船贸易”方式，与一些国家签订商约，当时中国的海船极其牢固和庞大，并且设施较为齐全，大批士兵随船往返，防止海盗的袭击。中国商船按期将生丝、花绸、缎、绢、金锦等运往各地，东起菲律宾及印尼各岛，西至印度的科泽科德、伊朗的霍尔木兹、伊拉克的巴士拉、也门的亚丁、沙特阿拉伯的麦加、埃及的杜姆亚特，直到大西洋滨摩洛哥的丹吉尔，南面可远销至马里的摩加迪沙、坦桑尼亚的基尔瓦等地。

The Overland Silk Road in the Yuan Dynasty came as far as the Black Sea and the Persian

Gulf due to its vast territory. However, the sea routes became more prosperous. Seven Shibo departments were set up in Quanzhou, Ningbo, Shanghai, Ganpu, Wenzhou, Guangzhou and Hangzhou to manage foreign trade. The ports in these cities were bustling with ships, especially the prosperous Quanzhou port, which impressed many foreign travelers. The Yuan Dynasty carried out foreign trade in the way of "Official Ship Trade" and signed agreements with some countries. China' s sea vessels were extremely strong and huge, and the facilities were relatively complete, with soldiers on board preventing pirate attacks. The merchant ships carried raw silk, flower noil poplin, satin, thin tough silk, and golden brocade to other countries, starting from the Philippines and Indonesian islands in the east to Kozhikode in India, Hormuz in Iran, Basra in Iraq, Aden in Yemen, Mecca in Saudi Arabia, and Damietta in Egypt on schedule, until Tangier in Morocco, Atlantic Ocean, and far south to Mogadishu in Mali and Kilwa in Tanzania.

明朝初期，由国家组成的大规模远洋航队为海外贸易的主要形式。1405 ~ 1433年，明朝郑和率领当时世界上庞大的、由300多艘宝船组成的舰队，穿过马六甲海峡，航行于印度洋，远达非洲东海岸。海外贸易的兴起，促进了苏州、杭州、漳州、潮州等地丝绸业的发展。明朝海上丝绸之路主要有广州、泉州、宁波三个主港和其他支线港。郑和完成了七次下西洋的壮举，带去了精美无比的丝绸、瓷器以及其他精湛的中华工艺制品，作为礼物和交换商品，其种类有湖丝、细绢、缎、丝绵、纱锦等四五十种。

The overseas trade was mainly secured by a large-scale state fleet at the beginning of the Ming Dynasty. From 1405 to 1433, Zheng He led the fleet of more than 300 ships in the world across the Straits of Malacca, sailing in the Indian Ocean, as far as the east coast of Africa. The rise of overseas trade boosted the development of the silk industry in Suzhou, Hangzhou, Zhangzhou, Chaozhou and other places, and Guangzhou, Quanzhou, Ningbo and other regional ports supported the sea routes in the Ming Dynasty. Zheng had achieved seven great missions, bringing abroad exquisite silk, porcelain and other delicate Chinese handicrafts. As gifts and commodities in exchange, there were about forty or fifty kinds of products, such as lake silk, finespun silk (with thin and tough texture), satin, silk brocade and yarn brocade.

明朝实施海禁，海上丝绸之路日渐衰落。海禁迫使民间海外贸易转型为以走私性质的私商贸易为主。民间海外贸易的需求张力和朝廷政策的矛盾冲突始终贯穿明清两朝。无政治武装支持的中国海商无力挑战大航海后政治军事商业合一的西方扩张势力，海禁导致中国退出海洋竞争。后来实行部分开海禁，康熙初年，外国商船获准在澳门停泊，与清政府特准的商人进行有限制的贸易。1685年，又设置江海、浙海、闽海、粤海四关，以粤海关为主要对外贸易港口。1762年又实行配售制。英国发明飞梭后，外商在中国主要以求购生丝作原料为主，丝织成品减少。

The Maritime Silk Road declined gradually in the Ming Dynasty due to the imposition of a sea ban, which forced conventional overseas trade into smuggled private one. The conflict between the needs of private overseas trade and imperial policy continued throughout the Ming and Qing Dynasties. Without the political support and armed forces to protect, the Chinese maritime merchants were unable to challenge the Western expansionary powers that integrated politics, military and commerce after the great voyage period, and the sea ban caused China to withdraw from the maritime competition Later, a partial ban was imposed. In the early years of Kangxi, foreign merchant ships were allowed to berth in Macau and conduct limited trade with merchants specially authorized by the Qing government. In 1685, the four ports of Jianghai, Zhejiang, Fujian and Guangdong were set up, with Guangdong as the main port of foreign trade. In 1762, the placement system was implemented again. After the invention of the flying shuttle in Britain, foreign businessmen mainly purchased raw silk as raw material in China, and the number of finished products decreased.

第三节 丝绸文化 /Silk Culture

丝绸文化具有5000 ~ 6000年的历史，是人类通过生产和使用丝绸活动所产生的物质和精神财富的综合。丝绸文化作为一种自身的文化，与中华传统文化的形成和发展息息相关。丝绸文化在中国诞生和发展，进入世界，并促进了与各国之间的政治、经济、文化、技术交流及其发展。

Silk culture has a history of 5,000 to 6,000 years, a combination of the material and spiritual wealth produced by mankind through the production and use of silk. It is closely related to the formation and development of traditional Chinese culture. Born and developed in China, it has promoted political, economic, cultural, and technological exchanges and their development between countries.

一、丝绸对经济的贡献 /The Contribution of Silk to Economy

丝绸产品，最初是为了满足人们的服装需求，逐渐成为产品，进而成为一种商品。这种商品交换日益频繁，促进了商业的繁荣。生产者不仅可以满足自己的需要，通过买卖还可以换取自己生活所需的其他商品，这就带动了经济的发展。丝绸作为一种商品，可以带动其他商品的流通，甚至以点带面，如明清时期的苏州，就是丝绸为代表的商品集散中心，阊门一带被誉为“天下第一码头”。

Originally aimed at satisfying people’ s fundamental life-sustaining needs, silk has gradually become products, a life necessity commodity. The increasing frequency of such commodity exchanges has contributed to the prosperity of commerce. Producers not only meet their own needs, but also buy and sell other goods needed for their own lives, which drives economic development. As a commodity, silk facilitates the circulation of other commodities. For example, Suzhou in the Ming and Qing Dynasties was a silk commodity distribution center, and the Lyumen area was known as "the first wharf in the world".

汉代布帛的价值计量功能突出，西汉时期，记录文字的载体由书简发展为将文字写在白色丝织物上，也就是帛书，重要的文件大多使用帛书。在魏晋南北朝时期，官员们都是用帛作为实物货币流通。隋唐时期，商品交换频繁，经济复苏，但布帛仍是法定流通货币。

The value measure function of cloth and silk in the Han Dynasty was prominent. In the Western Han Dynasty, the materials used to record words changed from booklets to white silk fabric, that is, silk manuscripts for important documents. In the Wei, Jin, Southern and Northern Dynasties, officials used silk for their money. The goods were exchanged frequently in the Sui and Tang Dynasties and the economy recovered, but cloth and silk were still the legal currency.

宋朝，丝绸商品经济已相当繁荣，丝织手工业和丝绸商业性城镇大量涌现，如开封、成都、苏州、杭州等。行业的相对集中又进一步推动了丝绸生产的商业化发展，并向周边地区扩散。随着北宋政权的南迁，我国丝绸业重心也随之南移。两宋时期，丝绸生产除官营以外，还有私营及官府民机包织等形式。同时，农村还出现了半脱离土地的准蚕织户，有些流入城市，成为待人雇佣的丝织工匠，还有些则拥机备料自织。

In the Song Dynasty, the silk commodity economy was quite prosperous, and many silk handicraft businesses and silk commercial towns emerged, such as Kaifeng, Chengdu, Suzhou, and Hangzhou. The relative concentration of the industry further promoted the commercialization of silk production and spread to surrounding areas. With the regime of the Northern Song Dynasty moving south, the silk industry also moved south. In the Song Dynasty, silk production was not only made by the government, but also by the private sector and with farm-out approaches. Besides, there were quasi-silk weaving households in rural areas that were semi-detached from the land. Some flowed into cities to become employed silk weaving craftsmen, while others owned machines to prepare materials for weaving.

明朝时期，江浙一带丝绸经济进一步发展，丝绸手工业及商业成为城市工业的主要产业。明朝中后期，苏州城内织工、染工各达数千人，东半城比户皆织，尚不包括城郭外五六十里地之农村，丝织业中已孕育出资本主义萌芽，出现稍具规模的织造工厂，设机督织，并分化出专业性较强的作坊主、机户和机匠。

In the Ming Dynasty, the silk economy in Jiangsu and Zhejiang developed further, and the silk handicraft industry and commerce became the main business in the urban industry. In the middle and late Ming Dynasty, there were thousands of weavers and dyers in Suzhou, and the eastern half of the city was all engaging in weaving, excluding the rural areas 50 or 60 miles outside the city. Capitalism had been born in the silk weaving industry, with the emergence of slightly large-scale weaving workshops equipped with machines and the differentiation of professional workshop owners, machine owners and machine artisans.

明末，苏南、浙北出现一批具有丝绸特色的新兴市镇，引来四方商贾，都到此收货，形成了零星上市和批量收购之间的矛盾，于是产生了专事丝绸贸易的牙行和牙人，市集牙行集零为整，转输他方，促进了丝绸贸易的进一步繁荣。

In the late Ming Dynasty, many towns with silk characteristics appeared in the south of Jiangsu and the north of Zhejiang, which attracted merchants from all over the country to receive goods here, resulting in the contradiction between sporadic listing and bulk purchase. As a result, there came into being some people and agencies who acted as mediators specializing in the silk trade. In the market, the agencies collected goods for wholesale and transferred them to other parties, which promoted the prosperity of the silk trade.

清康熙年间，社会趋于稳定，丝绸经济繁荣，行业内部分工细化，丝绸商贸活动活跃，丝绸生产的资本活动扩大，出现了颇具实力的牙行和包买主。与此同时，丝绸产区间的商旅活动和经济交流日益频繁，各帮客商来往于各丝绸市集间。为保障同业同行或同行利益，在求生存、同发展的意识下，丝绸各行业纷纷组织起来，成为自身的同行组织，中国丝绸行会逐渐步入发展期。至清朝中叶，中国丝绸行业组织已相当健全，成为全国工商行会组织中最具影响力的组织之一。

During the emperor Kangxi period of the Qing Dynasty, an era that witnessed a stable society, a booming silk economy, and refined division of labor within the industry, coupled with more silk trade transactions and investment capital of silk production. There emerged powerful agencies and buyers with wholesale purposes as well. Meanwhile, business trips and economic exchanges between the silk production areas had become increasingly frequent, and various merchants traveled between various silk markets. In order to protect the interests of the trade or peers, the silk industries had themselves organized and became their peer organizations under the consciousness of survival and development, with the silk guilds entering the development period. By the middle of the Qing Dynasty, the silk industry organizations had well developed, becoming the most influential ones in all the industrial and commercial associations.

以丝绸为代表的丝织工业后来成为中国最早出现资本主义萌芽的领域。中华人民共和国成立初期，以丝绸为代表的丝织工业是创汇、就业的主要来源，成为国民经济的中流砥柱。人民币、邮票、国庆巡游海报等上面均有丝织工人的形象。

The silk textile industry, represented by silk, later became the first area in China where capitalism sprang up. In the early days of the founding of the People' s Republic of China, the silk weaving industry was a major source of foreign exchange and employment, becoming the mainstay of the national economy. There were images of silk-weaving workers on the Chinese currency, stamps, National Day cruise posters and so on.

二、丝绸对文化的贡献 /The Contribution of Silk to Culture

丝绸是人类迈进文明社会的标志之一，它的出现揭开了人类的文明史，在整个人类文化的发展中处于特殊的地位。这种地位在形成文字较早的中国汉文化中，可以从词汇的形成过程看出其梗概。在汉语中存在大量与纺织和丝绸有关的文字、词汇、成语、诗歌及其他文学作品。在甲骨文、青铜器铭文中，已经发现了蚕、桑、丝、帛等的象形字体。在殷商的甲骨文中，以“系（纟）”为偏旁的字有100多个。以“系（纟）”为偏旁的文字中，关于丝绸纺织的有缫、绎、经、纬、绘、织、综、绳、纫、缪等；关于纺织品种类的有缯、纨、缟、绫、缎、绸、纱、绢、绡、绒等；关于纺织品色彩的有绿、缥、缇、绛、缁、缛等；关于服饰品的有缨、绅、绶、组、绔、绦等。随着时间的推移，到了南北朝时期，由纺织丝绸衍生出来的文字进一步增加。而在宋后流行的《大广益会玉篇》中则收“系（纟）”部459字，“巾”部172字，“衣”部294字。清朝，《康熙字典》中的“系（纟）”部约有830字，又较宋朝增加了很多。这充分说明了中国的语言文字与丝绸的关系是相当密切的。

The appearance of silk, one of the symbols of mankind’ s progress into civilized society, reveals the history of human civilization, whose status can be seen from the vocabulary in the early formation of Chinese Han culture. There are many characters, words, idioms, poems and other literary works related to textiles and silk in Chinese language. Pictographic fonts such as silkworm, mulberry, and silk have been discovered in Jiaguwen (inscriptions on bones or tortoise shells of the Shang Dynasty), and bronze inscriptions. In the Jiaguwen of the Yin and Shang Dynasty, there are more than 100 characters with 系（纟）as a part, among which there are about ten related to weaving techniques, ten to types of silk, six to colors of dyed silk, six to accessories. There are about 459 characters with 系（纟）part, 172 巾, 294 衣, according to a well-known book in the late Song Dynasty. In the Qing Dynasty, there were about 830 characters with 系（纟）as a part of the *Kangxi Dictionary*, which increases a lot compared with those of the Song Dynasty. This fully demonstrates a closer connection between the Chinese language and silk.

丝绸也影响诗歌和文学作品的创作。在我国最早的文学作品《诗经》中就有许多关于丝绸生产的描述，其中与丝绸文化相关的诗歌有二三十首，从不同的角度反映了春秋战国时期丰富的社会现实。有些诗歌反映了农业生产，如《七月》，它根据农事活动顺序，以平铺直叙的手法，逐月展示农业生产中的各个场景，其中蚕桑丝绸的生产过程得到了完整的反映。唐朝是我国诗歌创作的鼎盛时期，出现了许多与丝织相关的绝妙诗歌，众多著名的诗人都写下了这类诗篇或诗句，以丝织为题材最多的诗人要数白居易、王建等人。白居易一生作诗近三千首，数量之多在唐代诗人中首屈一指，大多是以反映劳动人民生活为题材的作品，其中有以丝织品为题材的诗篇，如《红绒毯》《缭绫》《重赋》《新制布裘》等。在中国的文学殿堂中，散文、小说、戏曲等对丝绸文化也有很多精彩的描绘，如著名的文学作品《红楼梦》等。蚕桑丝绸自从诞生之日起，就与文化艺术结下了不解之缘。丝弦乐器、翩翩舞蹈、书法绘画、帛书纸张，都与丝绸文化有关。在中国文化艺术殿堂中，含有丝绸文化的古典诗歌，以其喜闻乐见的形式、朗朗上口的韵律、比兴象征的手法、生动丰富的内容，更加显示出隽永和独特的魅力。

Silk also influences the composing of poetry and literature. There are many descriptions of silk production in the earliest Chinese literary work, the Book of Songs, among which 20 or 30 poems are related to silk culture, reflecting the social reality of the Spring and Autumn and Warring States Period from different perspectives. Some poems reflect agricultural production, such as *Qiyue* (*July*), which displays scenes of agricultural production month by month in accordance with the order of agricultural activities. The production process of mulberry silkworm silk is fully portrayed. The portrayal of wonderful poems related to silk weaving thrived in the Tang Dynasty, a peak period of poetry creation in ancient China. Among famous poets writing on the subject of silk are Bai Juyi, and Wang Jian. Bai wrote nearly 3,000 poems in his lifetime, the most productive Tang poet. Most of his poems reflect the lives of the grassroots, with some centering on silk weaving. There are also descriptions of silk culture in Chinese literary works, including essays, novels and operas, among which is the famous literary work *Hongloumeng* (*A Dream of Red Mansions*). Since its birth, silk has formed an inextricable bond with culture and art. All related to silk culture are the stringed instruments, ballast dance, calligraphy and painting, and silk manuscript paper. In the halls of Chinese culture and art, classical poetry with silk culture has even more meaningful and unique charm with its pleasing forms, catchy rhythms, symbolic techniques and well-presented contents.

作为中国画的载体，丝绸以优良特性对中国艺术人物画的风格和技法产生了很大的影响。中国画讲究神韵和主观感觉的表现，在技法上注重线条的灵活运用，所以在中国人物画中主要是通过线条描绘来表现人物的气质神韵，从而形成中国人物画的特点，即具有丝绸那种独特的静态悬垂感和动态飘逸感。宋朝时期的中国画除了以绸绢作为载体外，还出现了以绫锦作为画卷的装饰，即“装裱”，这是随着绘画艺术的发展而形成的一种特殊形式。采用绫锦做中国画的边框，其花纹和色彩可根据画的内容格调而选用，极富装饰性。不仅如此，采用绫锦作为中国画的边框也是绘画章法的一个组成部分，以满足中国画艺术构图的需要。中国画的构图，通常都是采用边框为依托的。另外，中国画绝大多数是卷绕的，将一幅很长的画卷卷绕成一个画轴便于收藏和携带，采用绫锦裱褙边框后，可以增加画卷的耐磨程度，便于长久保管收藏。

A vehicle for Chinese painting, silk has greatly influenced the style and techniques of Chinese artistic figure painting with its unique features. The painting pays attention to the expression of verve and subjective feeling and the flexible use of lines in techniques. Therefore, the figure painting, whose temperament and verve of characters are mainly depicted by lines, bears resemblance to what the silk looks like, such as a unique static sense of draping and a dynamic sense of elegance. Apart from that, the paintings in the Song Dynasty were decorated with silk brocade as the scroll, that is, framing, a special form shaped with the development of painting art. The pattern and color of the silk brocade are selected according to the content of the painting, which is a highly decorative method. Moreover, its use as the painting frame is also an integral part of painting rules, which meets the needs of Chinese painting' s artistic composition. In addition, most Chinese paintings are coiled. A long scroll is wound into a drawing shaft to facilitate collection and carrying. The use of a brocade frame improves the abrasion resistance of the scroll, facilitating long-term storage and collection.

三、丝绸对艺术的贡献/The Contribution of Silk to Art

中国的丝绸文化于20世纪30年代初第一次出现在银幕上，这与中国现代文学巨匠茅盾有着密不可分的关系。

Chinese silk culture first appeared on the screen in the early 1930s, which is closely related to Mao Dun, a giant in the history of modern Chinese literature.

1993年，电影《春蚕》问世（图1-4）。这部无声故事片由蔡叔声根据茅盾同名小说改编，是五四运动以来中国新文艺作品第一次被搬上银幕，为当时的中国电影带来了一种全

图1-4　电影《春蚕》
The Movie *Spring Silkworm*

新的内容，被称为“1933年影坛上的一个奇迹”。

The movie *Spring Silkworm* was on in 1933 (Figure 1–4), a silent feature film adapted by Cai Shusheng based on Mao Dun’ s novel of the same name. It was the first time that Chinese new literary works had been put on the screen since the May 4th Movement. It brought a new kind of content to Chinese movies, which was called a miracle in the film circle in 1933.

近年来，随着世界各地兴起学习汉语的热潮，中国国家汉语国际推广领导小组适时地推出了一批宣传中华文化的影视作品。其中，《中国文化精粹——丝绸》以话语解说和画面切入交相呼应的方式，展现了美轮美奂的中国丝绸文化。这个作品里面呈现了新石器时代以来不同时期的丝织精美产品、蚕丝绸生产工具、蚕形器物，沙漠、戈壁滩上骆驼商队行走于丝绸之路，郑和下西洋开辟的海上丝绸之路，现代女模特在轻歌曼舞中展示的丝绸时装秀。这个作品也展现出了许多蚕桑丝绸的历史发展、文化寓意、社会功能和在中西经济、文化交流过程中的重要意义，内容非常丰富。

In recent years, with the upsurge of learning Chinese all over the world, Hanban, the Confucius Institute Headquarters, has launched a batch of film and television works that promote Chinese culture in a timely manner. Among them, *The Core of Chinese Culture: Silk* shows the Chinese silk culture in the way of discourse interpretation and screen cutting. This work presents fine silk weaving products, silk production tools and silk worship artifacts from different periods since the Neolithic Age; the camel caravans walking on the Silk Road in the desert and the Gobi Desert, the maritime Silk Road opened by Zheng He’ s voyages to the Western seas, and the silk fashion show presented by modern female models in light songs and dances. It also shows, with rich content, many historical developments, cultural meanings, social functions and significance in the economic and cultural exchanges between China and the West.

由于蚕桑丝绸在国计民生中占有重要地位，一系列独特的生产习俗和风土人情在蚕桑地区形成。首先，各个蚕桑地区都形成了祭祀蚕神的传统风俗。历朝历代，皇宫内都设有先蚕坛，供皇后祭祀用，每当养蚕前，需杀一头牛祭祀蚕神，祭祀仪式十分隆重。其次，也形成了与蚕乡人民生老婚丧、衣食住行和精神生活息息相关的社交礼仪风俗。比如，在蚕乡社交方面，在采收蚕茧后，解除了“关蚕门”的禁忌，人们便开蚕门，走亲访友，互赠礼品，互问收成，俗称“望蚕讯”，体现了亲友的互相关爱之情。最后，办各种体现蚕桑丝绸文化的节庆集会。举办的各种类型的丝绸节，既弘扬了丝绸文化传统，又获得了巨大的经济效益。例如，在杭州西湖博览会期间，举办了杭州（国际）丝绸旅游文化节系列活动，为了弘扬丝绸之府、打造女装之都，推动杭州丝绸产业的提升与发展。还有中国苏州国际丝绸节的举办，不同于以往的丝绸搭台、商贸旅游唱戏，这次活动把丝绸节作为一个城市品牌来看待。这次活动的内容涵盖了丝绸文化、传统工艺、现代产业和优秀品牌展示等，展示产品种类丰富，传统工艺与现代成品相结合，让消费者对丝绸有了新的认识，是中国丝绸产业的一次集中巡礼。丝绸商铺的出现也大大促进了丝绸文化的推广与发展(图1-5)。

As sericulture silk occupies an important position in the national economy and people' s livelihood, a series of unique production practices and local customs have also emerged in some areas concerned. First, various sericulture areas have formed traditional customs of worshiping the god of silkworms. Throughout the dynasties, there have been silkworm altars in the imperial palace for the sacrifice of the queen. Before rearing silkworms, it is necessary to kill a head of cattle to sacrifice the silkworm gods, a solemn sacrifice ceremony. The worship of the gods is the most important activity in the customs of the sericulture countryside. Besides, social etiquette customs, which are closely related to the people of silkworm township, such as marriage and funeral, livelihood and spiritual life, have been formed. For example, in the social interaction of silkworm townships, after harvesting silkworm cocoons, the taboo of "closing doors during silkworm rearing" was lifted. After reopening their doors, people visited relatives and friends, gave gifts to each other, and asked each other about their harvest, commonly known as "looking forward silkworm-related news", reflecting the mutual greetings between relatives and friends. Finally, a variety of festive gatherings that reflect the silkworm mulberry silk culture were held. After the reform and opening up, various types of silk festivals were held in many places, which not only carried forward the silk cultural tradition, but also gained huge economic benefits. For example, during the Hangzhou West Lake Expo, a series of activities of the Hangzhou (International) Silk Tourism Culture Festival was held, in order to promote the city of silk, and women' s clothing, advancing the development of Hangzhou' s silk industry. Different from the past festival of its kind, where trade and tourism are the focus instead of silk, the Suzhou International Silk Festival

图1-5 丝绸商铺
Silk Shops

sees the festival as a city brand. It covers the contents of silk culture, traditional craftsmanship, modern industries and outstanding brand displays, showcasing a wide variety of products, and integrating traditional craftsmanship with modern finished products, giving consumers a new concept of silk, an important window into China' s silk industry. The emergence of silk shops has also greatly promoted the development of silk culture (Figure 1–5).

这些丝绸节各具特色，形式不尽相同，板块组合各异，但是它们有着共同的目标，这就是：①弘扬中国丝绸文化；②以文化支持民族品牌的发展；③营造消费者与商家、企业互动、产供销信息互通的市场环境；④提高人们对中国丝绸精神与物质、文化与品牌的认知；⑤提升丝绸品牌形象，带动丝绸消费市场，激发企业创新意识；⑥适应国内外市场需求，以市场促进品牌建设，实现产业发展、市场繁荣和百姓受益的良好局面。

These festivals have their characteristics, with different forms and modules, but they share the same objectives of achieving industrial development, market prosperity and the benefit of the people: ① to carry forward Chinese silk culture; ② to support the development of national brands with culture; ③ to create a market environment in which consumers interact with merchants, enterprises, production, supply and marketing information; ④ to raise people' s awareness of the Chinese silk culture and brands, spiritually and materially; ⑤ to promote the image of silk brands, the consumer market, and the enterprise innovation; and ⑥ to accommodate the changes of the domestic and foreign markets, with which to promote brand development.

第二章 蚕桑与制丝技艺
Silkworms and Silk Making Techniques

我国是世界上最早开始种桑养蚕的国家，至今约有5500年的历史。目前，世界上共有50多个国家饲育桑蚕，生产桑蚕茧，主要有中国、印度、朝鲜、越南和巴西等国。中国产量最高，印度次之。我国主要产区有广东、广西、浙江、江苏、重庆、四川等。

China, the first country in the world to start planting mulberry and rearing silkworms, has a history of about 5,500 years. Rearing silkworm and producing silkworm cocoons are available more than fifty countries in the world, mainly including China, India, North Korea, Vietnam and Brazil, among which China has the highest yield, followed by India. The main production areas in China are Guangdong, Guangxi, Zhejiang, Jiangsu, Chongqing, Sichuan and others.

第一节 栽桑与养蚕技艺/Mulberry Planting and Silkworm Rearing Techniques

一、栽桑技艺/Mulberry Planting Techniques

我国幅员辽阔，生态环境各异，经过长期的自然选择和人工选育下，产生了许多地方品种。我国长江三角洲、珠江三角洲和四川盆地的自然条件适宜栽桑养蚕。桑树主要有四大品种：鲁桑（山东）、白桑（新疆）、山桑（山西格鲁桑）和广东桑（广东）。

Many local varieties emerged under long-term natural selection and artificial breeding in China, one with a vast territory and different ecological environments. Most suitable for planting mulberry and rearing silkworms are the Yangtze River Delta, Pearl River Delta and Sichuan Basin. There are four main varieties of mulberry: Lusang (Shandong), Baisang (Xinjiang), Shansang (Gelusang in Shanxi) and Guangdongsang (Guangdong).

早在商周时期，随着养蚕业的迅猛发展，野生桑树已不能满足需要，于是开始进行人工栽桑。桑树的栽培分为有性繁殖和无性繁殖两种。有性繁殖就是用桑籽播种育成实生苗，一般用来作为嫁接用砧木；而无性繁殖则是利用枝条或桑树根进行嫁接、扦插、压条来繁殖。蚕桑的优良苗木繁殖一般都采用无性繁殖。

As early as the Shang and Zhou Dynasties, with the rapid development of the sericulture industry, wild mulberry trees could no longer meet the needs, so the domestication of mulberry trees began. Mulberry cultivation is divided into sexual and asexual reproduction: the former refers to sowing mulberry seeds to breed solid seedlings, which are commonly used as grafting roots; and the latter using branches or mulberry roots for grafting, cutting and pressing, one that is generally used for good mulberry seedlings.

到了西周时期，人们认识到桑树适宜种植在高燥的地方，如在《诗经》一书中有这样的记载："南山有桑"和"无逾我墙，无折我树桑。"这里的"南山"的"山"是指山坡，而"无逾我墙，无折我树桑"则是指住宅里的桑树，而古代的住宅大多建在高地上。此时，人们还懂得采用撒籽播种的方法来繁殖桑树。而在西周以后，桑枝修整的技术更为成熟，如《诗经・七月》云："蚕月条桑，取彼斧斨，以伐远扬，猗彼女桑。"意思是说，三月里修整桑条，拿起砍柴刀，把太长的枝儿全砍掉，用嫩枝儿扎牢。

In the Western Zhou Dynasty, people realized that mulberry trees were suitable for planting in dry places. For example, *The Book of Songs* carries the description of "There are mulberry trees in the south mountain", and "Don' t climb over my fence, and break my mulberry tree. " The "mountain" here refers to the hillside, and the latter description refers to the mulberry tree planted in a yard of a house built on the highlands. It was a time when people knew how to breed mulberry trees by means of seed sowing. The techniques of trimming mulberry branches were maturer after the Western Zhou Dynasty, as expressed in *The Book of Songs (July)* "Trimming the mulberries in March, picking up the machete, cutting off all the branches that are too long, and fastening them with tender branches. "

在秦汉时期，栽桑养蚕技术得到了进一步发展。我国最早关于农业的书籍是西汉的《氾胜之书》，书中详细地记载了桑树的栽培方法。由于桑树修整技术的不断提高，树的形态也不断发生变化，由原来的自然形变化成高干、中干和低干三种类型，修剪形式也由原来的"无拳式"发展到"有拳式"。桑树的修枝整形是我国古代劳动人民的独特创造，可以提高桑叶的产量和质量，以满足养蚕业发展的需要。

During the Qin and Han Dynasties, the technique of planting mulberry and rearing silkworms developed further. The earliest farming book in China is *The Book of Fansheng* in the Western Han Dynasty, which details the cultivation methods of mulberry trees. Due to the continuous improvement of the mulberry tree pruning technique, the shape of the tree had also been continuously changed, from the original natural deformation to three types: high dryness, middle dryness and low dryness, and the pruning form had also developed from the original to a new one. The pruning and shaping of mulberry trees is a unique creation of the ancient hardworking Chinese, which can improve the yield and quality of mulberry leaves to meet the needs of the development of the sericulture industry.

在隋唐时期，桑蚕技术没有新的变化，直至宋元时期，桑蚕技术才有所创新。从北宋开始，我国南方的蚕农发明了先进的桑苗嫁接技术，这对老桑树的复壮更新、保持桑树优良性状、加速桑苗的繁殖以及培育优良品种都具有十分重要的意义。由于我国幅员辽阔，不同地区培育的桑树品种也有较大的差异。例如，我国华南地区以地桑为主，华北地区以中、高干桑为主，华东地区则以低干桑为主。

During the Sui and Tang Dynasties, there was no change in the relevant techniques, which had not been innovated until the Song and Yuan Dynasties. From the Northern Song Dynasty, silkworm rearers in southern China developed advanced techniques for grafting mulberry seedlings, which are of great significance for the growth and renewal of old mulberry trees, maintaining their excellent characteristics, accelerating the reproduction of mulberry seedlings and cultivating excellent varieties. The varieties of mulberry trees cultivated in different regions are also quite different due to China’ s vast territory. For example, according to the height of the tree trunk, South China is dominated by mulberry planted with proso millet, North by the medium and high trunk, and East by the low trunk.

桑树是经济价值很高的物种（图2–1），桑叶是家蚕唯一的饲料（图2–2）。在正常情况下，春季桑树的新梢每3 ~ 4天就长出一片新叶，叶子从开叶到成熟需20 ~ 25天。由于桑叶富含生物碱成分，可用来制成茶叶，作为抗癌的保健品。桑葚是桑树的果实，在古代曾作为皇室的补品。桑皮可入中药，有止血、润肺、止咳的功效，还有淡化疤痕、利水消肿之功效，也可用来加工成纺织纤维，用于纺纱织布制衣裳。桑树的嫩枝条可以用来酿酒或入药，也可用作造纸的原料。

Mulberry is a species of high economic value (Figure 2–1), and mulberry leaves are the only feed for silkworms (Figure 2–2). Under normal conditions, the new shoots of mulberry trees in spring grow one new leaf every three to four days, and the leaves take twenty to twenty–five days from leaf opening to maturity. Rich in alkaloids, mulberry leaves can also be used to make tea as anti–cancer healthcare products. Mulberries are the fruit of a mulberry tree, which used to be a supplement to the royal family in ancient times. Mulberry peel can be used in traditional Chinese medicine to stop bleeding, moisten the lungs and relieve cough, to desalinate scars and reduce swelling, and to make textile fibers for spinning and weaving clothes. The tender branches of mulberry trees can be used for brewing wine or medicine and as raw materials for paper.

图2-1　桑树
Mulberry

图2-2　桑叶
Mulberry Leaves

二、养蚕技艺 /Silkworm Rearing Techniques

1. 蚕的种类 /Species of Silkworms

蚕的品种很多，可分为家蚕（即桑蚕）和野蚕两类。家蚕以桑叶为饲料，故其丝称桑蚕丝，它是天然丝的主要品种，并且其质量好，俗称真丝或厂丝。野蚕的品种很多，有柞蚕、蓖麻蚕、樗蚕、樟蚕、天蚕及柳蚕等。野蚕有的可在室外放养，所食饲料也因蚕种的不同而不同，其中以在柞树上放养的柞蚕为主。然而，除柞蚕和天蚕以外，其他品种的野蚕茧均不能缫丝，一般将它们切成短纤维作绢纺的原料。

There are many species of silkworms, which can be divided into two categories: domestic silkworm (mulberry silkworm) and wild silkworm. With mulberry leaves as feed, domestic silkworms produce mulberry silk, which is the main source of natural silk and has the best quality. There are many species of wild silkworms, some of which can be kept outdoors, whose feed is different depending on the specific species. The commonly seen species is tussah silkworms stocked on tussah trees. With the exception of tussah silkworm and Japanese tusser, other varieties of wild silkworm cocoons cannot be reeled, and they are usually cut into short fibers as raw materials for tussah spinning.

家蚕（图2-3）属节肢动物门、昆虫纲、鳞翅目、蚕蛾科、家蚕种，起源于野外桑树上食桑的野蚕。桑蚕有中国种、日本种及欧洲种三个品系。桑蚕所结的茧称为桑蚕茧（图2-4），又称家蚕茧，由它缫得的丝称为桑蚕丝。

Silkworm, also known as a domestic silkworm or bombyx mori (Figure 2-3), belongs to arthropod, insecta, lepidoptera, and bombycidae, originating from wild silkworms that eat mulberries on mulberry trees in the wild. It has three species: Chinese, Japanese and European. The cocoon formed by the silkworm is called the silkworm cocoon (Figure 2-4), also known as the domestic silkworm cocoon, from which mulberry silkworm silk is obtained.

图2-3　桑蚕
Mulberry Silkworm

图2-4　桑蚕茧
Mulberry Silkworm Cocoon

柞蚕属节肢动物门、昆虫纲、鳞翅目、蚕蛾科、蚕蛾属、柞蚕种（图2-5）。柞蚕所结的茧称柞蚕茧（图2-6）。由柞蚕茧所缫制的丝称柞蚕丝，一般用于织造中厚型丝织品。柞蚕有中国种、印度种及日本种三个品系，它生长在野外的柞树（即栎树）上。我国的柞蚕品种很多，目前放养的主要有一化性和二化性两种，前者一年放养一次，所结的茧既作种茧，又是缫丝原料；后者一年放养两次，春茧只作秋茧的种茧，秋茧作为缫丝原料。

The tussah belongs to bombyx mandarina (Figure 2-5). The cocoon from the tussah is called the tussah cocoon (Figure 2-6), from which the tussah silk is made. The silk is generally used for weaving medium and thick silk fabrics. There are three species of tussah: Chinese, Indian and Japanese. It grows on oak trees in the wild. There are many varieties of cocoon silkworms in China. At present, there are two main types of cocoon silkworms: univoltine and bivoltine. The former means rearing silkworms in spring, with their cocoons for egg production and as raw silk for reeling, the latter in spring for reproduction and in autumn for reeling.

图2-5　柞蚕
Tussah

图2-6　柞蚕茧
Tussah Cocoon

蓖麻蚕（图2-7），属节肢动物门、昆虫纲、鳞翅目、天蚕蛾科。它原是野外生长的野蚕，食蓖麻叶，也食木薯叶、鹤木叶、臭椿叶、马桑叶及山乌桕叶，是一种适应性很强的多食性蚕。蚕和蚕茧通常是按所食叶类来命名的，如蓖麻蚕和蓖麻蚕茧，木薯蚕和木薯蚕茧等。现在的蓖麻蚕大多生长在野外，由人工放养，也有在室内由人工喂养的。

The castor silkworm (Figure 2–7) belongs to the arthropod, insecta, lepidoptera, and saturniidae. The Castor silkworm, originally a wild silkworm, eats castor and cassava leaves, crane leaves, odor leaves, and cedar leaves, an adaptable multi–eating silkworm. Silkworms and cocoons are usually named after the leaves they eat, such as castor silkworm and castor cocoons, cassava silkworm and cassava cocoons. Most of today' s castor silkworms live in the wild and are reared for human use, with some fed indoors.

天蚕（图2–8），属节肢动物门、昆虫纲、鳞翅目、天蚕属、天蚕种。它是一种生活在天然柞林中的一化性四眠五龄完全变态的昆虫，以卵越冬。其幼虫的形态与柞蚕相似，只能通过柞蚕幼虫头部有黑斑而天蚕没有这一点来加以区别。天蚕幼虫体呈绿色，多瘤状突起。食山毛榉科栎属树叶，如柞、赤栎、橡、白栎、槲树叶等。天蚕丝珍稀，价格昂贵，高于桑蚕丝、柞蚕丝近百倍，经济效益令人咋舌。

The wild silkworm (Figure 2–8) belongs to arthropod, insecta, and lepidoptera. It is a univoltine insect that lives in a natural oak forest. The shape of its larvae is very similar to that of the tussah silkworm, which can only be distinguished by the black spots on the heads of the tussah silkworm larvae, which the wild silkworm does not have. Its larva is green and has a multi–cancerous protuberance. The silkworm eats tree leaves, such as oak, red oak, white oak, and live oak. Its silk is rare and costly, nearly 100 times higher than that of mulberry silk and tussah silk, and its profit is staggering.

图2–7　蓖麻蚕
Castor Silkworm

图2–8　天蚕
Wild Silkworm

2. 蚕茧的形成 /The Formation of a Cocoon

蚕吐丝结的茧由茧衣、茧层、蛹体和蛹衬四部分组成。茧衣是茧子表层的类似棉状的茧丝，是蚕在簇中寻找合适结茧场所及构成茧子轮廓时吐出的凌乱疏松的丝。茧衣结成后，茧在茧衣内继续吐丝，此时蚕的头部摆动和身体移动都有规律，使吐出的丝规则地排列起来，每吐15 ~ 25个丝圈形成一个茧片，再转移位置继续吐丝形成另一个茧片，如此连续吐丝，最终茧片层层重叠形成茧层。当吐丝临近终了时，蚕体缩小，吐丝量明显减少，速度也逐渐变慢，此时头部的摆动已无规律，因此丝缕的重叠比较紊乱，形成松软的薄层，这是用来保护蛹体的衬垫，称蛹衬或蛹衣。其中，茧衣和蛹衬不能用于缫丝，只能用作绢纺原料，只有茧层可用来缫丝、织造丝绸。吐丝结茧完成后，蚕就在茧内蜕皮化蛹。初期的蛹体呈乳白色，体皮柔嫩。随着蛹的发育，体色渐渐变为淡黄、黄色、黄褐色，最后变成茶褐色。

The cocoon spun by silkworm is composed of four parts: cocoon coat, cocoon layer, pupal body and pupal lining. The cocoon coat is a cotton-like cocoon silk on the surface of the cocoon. It is a messy and loose silk that silkworms look for suitable cocooning site and form a cocoon contour. After the cocoon outer floss is formed, the silkworm continues to spin in the cocoon coat. At this time, the head of the silkworm swings and the body moves regularly, so that the silk can be arranged regularly, with each cocoon piece formed by every fifteen to twenty-five silk coils. The silkworm then shifts its position and continues to spin to form another piece, followed by continuous spinning, and eventually, the pieces overlap to form cocoon layers. When the spinning nears to the end, the silkworm body shrinks, the spinning amount obviously decreases, and the speed gradually slows down. At this time, the head swinging irregularly, the overlap of the silk threads is disorderly and a thin, soft layer emerges, which is used to protect pupal cells, called basin residue. Cocoon outer floss and basin residue can not be used for reeling silk, but for silk weaving materials; instead, the cocoon layer can be used for reeling and weaving silk. After finishing a cocoon, the silkworm molts and pupates inside the cocoon. The pupa is milky white in the early stage and its skin is tender. With the growth of pupa, the body color gradually becomes yellowish, yellow and brown, and finally becomes tea brown.

用作缫丝的原料茧由鲜茧干燥后得到。蚕茧的干燥过程称为“烘茧”，这是蚕茧加工的第一道工序，通常是利用热能将茧内的活蛹杀死，并除去适量的水分，便于保全丝的品质和长期储存。

Raw cocoons for silk reeling are obtained by drying fresh cocoons. The drying process of cocoons is called the cocoon drying, the first step of cocoon processing, in which the live pupa in cocoon is usually killed by heat and a proper amount of moisture is removed, facilitating the preservation of silkiness and long-term storage.

茧的外观性状是鉴定茧质量的重要依据，与制丝工艺也有密切的关系。茧的外观性状是指茧的形状和大小、茧的颜色和光泽等，它直接影响蚕丝的质量。茧的形状一般有圆形、椭圆形、束腰形、尖头形和纺锤形等。我国所产的蚕茧多为圆形、椭圆形和尖头形。茧的颜色一般为白色和黄色，也有少数为淡绿色、粉红色、乳黄色等。优质茧通常颜色洁白，光泽正常。色泽不一的蚕茧不仅舒解困难，而且容易产生夹花丝，影响蚕丝的质量。

An important basis for the identification of quality, the appearance of the cocoon, including shape, size, color and luster, is closely related to the spinning process. The shape of the cocoon is generally round, elliptical, peanut shaped, and spindle-shaped. Most of the cocoons produced in China are round, oval and pointed. The color is generally white and yellow, and a few are light green, pink, and creamy yellow. High-quality cocoons are usually white in color and gloss normally. The cocoons with different colors are not only hard to be unknotted, but prone to produce knots, which affects the quality of the filaments.

第二节 制丝技艺 /Silk Thread Making Techniques

蚕茧通过制丝工艺成为蚕丝，才能用于织造丝绸。制丝工艺包括混茧、剥茧、选茧、煮茧、缫丝、复摇、整理、包装以及检验。简单来说，混茧就是把从各地收购来的蚕茧进行混合；剥茧就是剥去蚕茧上的茧衣；选茧就是对原料茧进行分类分级，即分为上茧、次茧和下茧三类；煮茧就是把蚕茧煮成熟茧供缫丝用；缫丝就是将煮熟的茧索出绪丝，按设计规格和品位缫成生丝；复摇就是将缫丝得到的小篗丝重新卷绕成大篗丝片或摇到筒管上，即加工成筒装丝；整理就是将复摇后的大篗丝片进行平衡和编丝成绞，即加工成绞装丝；包装就是将筒装丝或绞装丝打包；检验就是根据国家标准，确定生丝等级。

The cocoon becomes silk through the process of making silk before it can be used to weave silk. The silk making process includes cocoon mixing, cocoon floss stripping, assorting, cooking, reeling silk, rereeling silk, finishing, packaging and testing. To put it simply, cocoon mixing refers to the mixture of cocoons acquired from different places; cocoon floss stripping to peeling off the cocoons; cocoon assorting to the classification of the raw cocoons according to their quality: excellent, moderate, and poor; cocoon cooking to making it boiled enough for reeling; reeling silk to getting raw silk from boiled cocoons according to specifications and quality; rereeling silk to the process of rewinding the reeled small winding silk into large ones or onto a round object to make the silk in the bobbin; finishing to the process of making the rereeled large winding silk in skein; packaging to packing the raw silk in bobbin and skein; testing to determining the grade of raw silk according to national standards.

一、混茧与剥茧 /Cocoon Mixing and Floss Stripping

在丝绸织造之前，首先需要准备原料蚕丝，而混茧和剥茧就是其中第一步。一般情况下，一个蚕庄的茧产量并不能满足大批量生产的要求，因此必须先进行混茧，混茧要求茧的质量相同，且各庄需要按照一定比例进行混茧，另外春秋茧不相混、新旧茧不相混。剥茧就是将茧层外面浮松的不能缫丝的茧衣剥除掉。剥去茧衣的光茧便于识别蚕茧的质量，且易煮熟均匀、鉴别煮熟程度，因此也有利于煮茧和缫丝。目前的剥茧机分为人工剥茧机和机械剥茧机，人工剥茧机产量在100～200千克/小时，如D031A、F223型设备等，机械剥茧机产量最高可达600千克/小时，如ZD102型。在剥茧过程中，需要注意给茧时要薄和匀，剥茧后可以通过剥光率（500粒茧中的全毛茧和半毛茧粒数）来检查剥茧质量。

Before weaving, the raw silk must be prepared, and cocoon mixing and floss stripping are the first step. Generally, the cocoon output of a silkworm house cannot meet the demands of mass production, so cocoon mixing must be carried out first, which requires the same cocoon quality. Each silkworm house needs to mix cocoons according to a certain proportion. In addition, spring and autumn cocoons are not mixed, nor the new and old ones. Cocoon floss stripping is to peel off the outer floss that is loose outside the cocoon layer and cannot be reeled. The stripped cocoon helps to examine its quality, boil uniformly and determine the degree of cooking; thus, it is also beneficial to cooking cocoons and reeling silk. There are currently manual and mechanical cocoon floss stripping machines. The former has an output of 100–200 kg/h, such as the equipment D031A, F223, and the latter of up to 600 kg/h, such as ZD102. In the process of cocoon peeling, it is necessary to make sure that the cocoon feeding is thin and uniform. After the cocoon peeling, the quality can be checked by the peeling rate (number of granules of all hairy cocoons and half-hairy cocoons in 500 cocoons).

二、选茧/Cocoon Assorting

选茧就是将蚕茧按其质量进行分类，需要分清上茧、次茧和下茧，然后按质量和种类对上茧进行分级和分型，目的是合理使用原料，将高质量的茧合并生产较高品质的生丝，同时通过分级和分型，可以给煮茧和缫丝创造有利条件。上茧又称头号茧，茧形、茧色、茧层厚薄等均较正常，无明显疵点。有时需对头号茧进行精选，头号茧的茧幅整齐率在80%以上，二号茧的茧幅整齐率在60%～80%，三号茧的茧幅整齐率在60%以下，三种等级的茧形与茧色也有所差异。此外，在缫丝之前，需对统号茧进行分型然后上车，一般分为大型茧（春茧19毫米以上，夏秋茧17毫米以上），中型茧（春茧17～19毫米，夏秋茧14～17毫米）和小型茧（春茧17毫米以下，夏秋茧14毫米以下）。次茧是有明显瑕疵但不到下茧标准的茧，而下茧则包括双宫茧、黄斑茧、柴印茧、畸形茧等。目前选茧一般采用传送带选茧机进行人工筛选，选茧机型号包括D031A、XJ73，产量为30～60千克/小时。为增加产量提高效率，目前通常采用混、剥、选茧输送连续化机组进行生产，如图2–9所示。

The process of cocoon assorting is to classify cocoons into three grades according to quality, and further group the excellent cocoons according to their quality and type. The purpose is to rationally use raw silk, gathering quality cocoons to produce high-grade raw silk, and creating favorable conditions for cocoon cooking and silk reeling. Without obvious defects, the excellent cocoon, known as the first-rank cocoon, is characterized by high standards in shape, color and layer thickness. Sometimes it is necessary to select the first-rank cocoon, whose integrity rate is over 80%, followed by second-rank 60% to 80%, and third-rank below 60%. All these groups vary in terms of color and shape. In addition, it is necessary to type the first-rank cocoons before reeling silk, and then get them reeled. They are generally divided into three types: large (spring cocoons above 19 mm, summer-autumn 17 mm), medium (spring cocoons 17-19 mm, summer-autumn14-17 mm) and small (spring cocoons below 17 mm, summer-autumn below 14 mm). Second-rank cocoons have obvious defects but are better than the third-rank ones, which include dupion, yellow spotted cocoon, cocoon pressed by cocooning frame, and malformed cocoon. At present, cocoon assorting generally uses conveyor belt cocoon assorting machines for manual screening. The machines include D031A and XJ73, and the output is about 30-60 kg/h. In order to increase the efficiency, a continuous production line is adopted that integrates cocoon mixing, floss stripping and assorting, as shown in Figure 2-9.

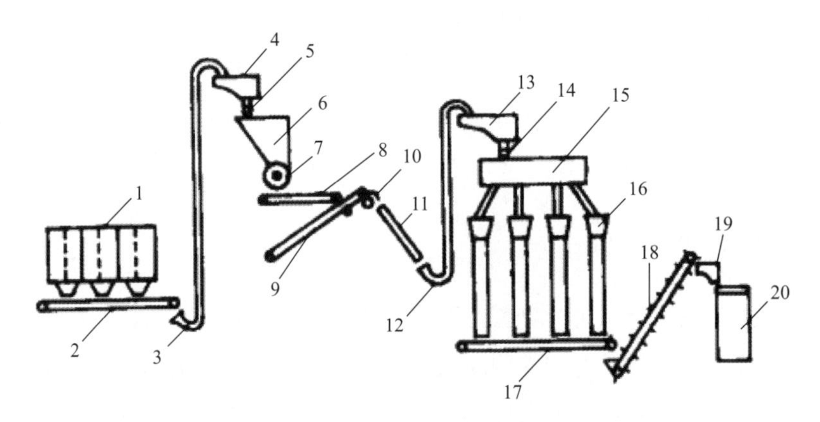

图2-9　混、剥、选茧输送连续化机组

A Continuous Production Line Integrating Cocoon Mixing, Floss Stripping and Assorting

1—混茧机/Cocoon mixing machine　2—毛茧输茧带/Cocoon conveyor belt　3—毛茧进料口/Cocoon inlet　4—毛茧沉降室/Cocoon settling chamber　5—毛茧卸料斗/Cocoon discharge hopper　6—毛茧箱/Cocoon box　7—抓茧辊筒/Cocoon roller　8—铺茧带/Cocoon belt　9—剥茧带/Cocoon stripping belt　10—刮刀/Scraper blade　11—接茧板/Cocoon transfer board　12—光茧进料器/Light cocoon feeder　13—光茧沉降室/Light cocoon deposit chamber　14—光茧卸料器/Light cocoon unloader　15—光茧箱/Light cocoon box　16—选茧台/Cocoon selection table　17—集茧输送带/Cocoon conveyor belt　18—上袋机/Bagging machine　19—自动计量器/Automatic measurement　20—茧袋/Cocoon packaging

在选茧过程中，为确保选茧质量，需建立各类茧、各级茧的标样，在选茧过后，可通过误选率（随机取样一包，数出200粒中的次茧和下茧，计算误选率，下茧中不可有上茧）评价选茧质量。若进行茧分型则采用筛茧机进行大小的筛选，筛茧机型号包括KC241、ZDS242，在筛选春茧和秋茧时产量为100～300千克/小时。

In the process of cocoon assorting, cocoon samples at all levels must be selected to ensure the quality, and then the effects can be evaluated by a false selection rate. If the cocoon classification is performed, the size screening is carried out by using a cocoon screening machine. The cocoon screening machine includes KC241 and ZDS242, and the output ranges from 100 to 300 kg/h when screening spring and autumn cocoons.

三、煮茧/Cocoon Boiling

选茧后，需要利用水、热能等作用，破坏茧层丝缕之间的胶着状态，使茧层丝胶适当膨润、软化和溶解，以便抽出绪丝来进行后续的缫丝工序，这个过程称为煮茧。在煮茧的过程中，要使水分均匀渗透进各茧层，这是茧层膨润的条件，渗透过程一般在煮茧机的渗透区完成，在高温蒸汽阶段（98～100℃）运行1～3分钟，使茧腔内的空气升温膨胀而排出腔外。然后进入低温煎煮阶段（50～70℃），利用内外温度差使茧层吸水膨润，但此阶段不可煮熟。随后经渗透的茧再经过煮熟，使丝胶膨润，以便在缫丝过程中能顺次解离。在煮熟过程中，相比水煮，蒸煮形式可使茧层各部位受热更均匀，一般保持在98～100℃，茧子进行快速吐水，完成煮熟茧的目的，期间压力也保持稳定，煮茧汤pH保持在7左右，酸度在10以内，目前一般要求煮茧的茧层丝胶溶失率在5%左右。在煮熟过程中需要根据实际情况进行调整，使煮熟程度更加均匀一致，包括调整温度、浓度以及调整部的动摇。有些工厂在煮茧前进行触蒸处理，尤其对洁净和解舒较差的茧，一般采用高温98～100℃的蒸汽经过10～15分钟的触蒸前处理，充分冷却24小时以上再进行煮茧。常用煮茧机为循环式蒸汽煮茧机，包括单蒸型循环式蒸汽煮茧机（图2-10）和汤蒸型循环式蒸煮机（图2-11）等，其工艺过程基本相同。

After assorting, it is necessary to make sericin softened and ultimately dissolved between the silk strands by water and heat, to facilitate the withdrawal of thread silk for the subsequent reeling process, which is called cocoon cooking. During cocoon boiling, water shall be made to penetrate each cocoon layer evenly, which is the condition for the cocoon layer to expand. Generally, the penetration process shall be completed in the penetration zone of the boiling machine and be operated for 1 to 3 minutes at the high-temperature phase (98-100℃), so that the air in the cocoon cavity could be heated up, expanded and discharged. This is followed by the low-temperature decoction phase (50-70℃), at which the cocoon layer absorbs water and swells by the internal and external temperature differences, but it cannot be boiled. The permeated cocoons are then cooked to make the sericin swell, so that it can be successively dissociated in the reeling process. In the cooking process, all parts of the cocoon layer are more evenly heated, with temperature kept at 98 to 100℃. The guttation of the cocoon can be quickly finished, the pressure stable, the pH kept at about 7 and the acidity within 10. At present, the cocoon layer sericin dissolution rate is generally required to be about 5%. The cooking process needs to be adjusted according to the actual conditions, making the cooking degree more uniform, including the temperature, concentration and changes of the adjusted part. Before cooking cocoons, steam jet treatment is adopted by some factories, especially the treatment of the cocoons that lack cleanness and are hard to be unknotted. Generally, steam with a high temperature of 98 to 100℃ is used for 10–15 minutes pretreatment. The cocoons can be fully cooled for more than 24 hours before cooking cocoons. The commonly used machine is a circulating steam cocoon cooking machine, including single steamed machines (Figure 2–10) and bath steamed ones (Figure 2–11). The process is the same.

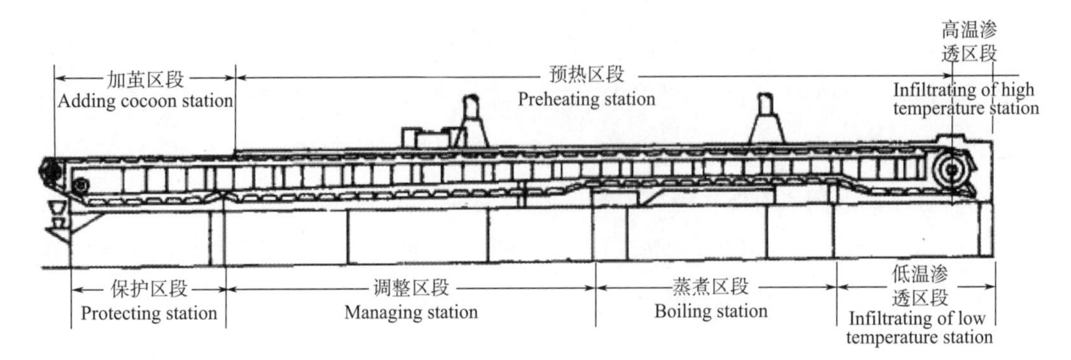

图2–10 单蒸型循环式蒸汽煮茧机
Single Steamed Circulating Cooking Machine

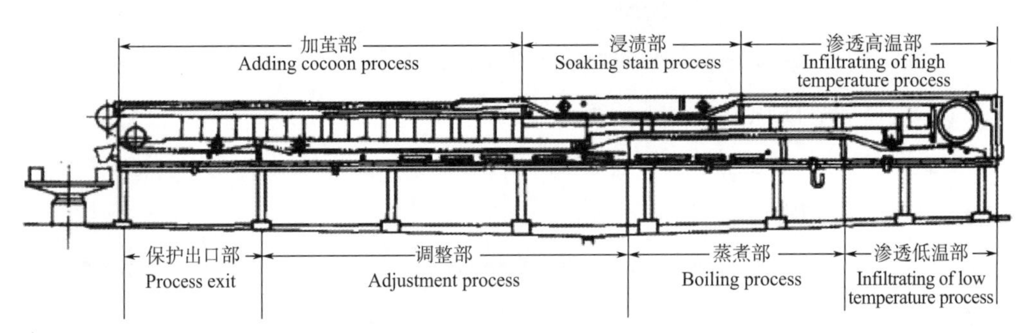

图2–11 汤蒸型循环式蒸煮机
Bath Steamed Circulating Cooking Machine

鉴别煮熟茧的方式：①适熟茧颜色呈白色或水玉色，茧层软滑有弹性，绪丝适量易拉出，茧蛹体稍硬，吸水率在95% ~ 97%，丝胶溶失率在4% ~ 6%，绪丝量在16 ~ 22毫克；②偏熟茧为水灰色或微黄色，茧层软而无弹性，绪丝脆弱易断，蛹体软而膨大，可以挤出水，吸水率在97%以上，丝胶溶失率在6%以上，绪丝量在22毫克以上；③偏生茧接近洁白色，与生茧相近，茧层粗糙而硬，绪丝少不易引出，蛹体硬且腹部不成形，吸水率在95%以下，丝胶溶失率在4%以下，绪丝量在16毫克以下。

The methods for identifying cooked cocoons are as follows: ① for normally boiled cocoons, the cocoon color is white or water jade, the layer soft and elastic, the floss easy to pull, the pupa slightly hard, and water absorption rate 95% to 97%, dissolution rate of sericin between 4% and 6%, amount of floss between 16 and 22 milligrams; ② for well-boiled ones, color gray or slightly yellow, lay soft and inelastic, floss fragile and easy to break, pupa soft and big, water absorption rate above 97%, the dissolution rate of sericin above 6%, amount of floss above 22; and ③ for less boiled ones, a color close to white and similar to raw cocoons, layer rough and hard, floss fewer and hard to pull, pupa hard and with abnormal belly, water absorption rate below 95%, the dissolution rate of sericin below 4%, amount of floss below 16.

四、缫丝/Reeling

煮茧之后就是缫丝工序，将蚕丝从蚕茧中抽出，其目的是：①将煮熟茧通过索理绪找到正绪，将数根茧丝合并；②通过集绪和丝鞘的作用形成丝条；③通过卷装和干燥作用形成小卷丝片。在缫丝过程中，要注意控制缫折。目前广泛采用的是立缫机和自动缫丝机，而自动缫丝机（图2-12）是在立缫机基础上发展起来的，目前占比也较小，其工艺过程与立缫机基本相同，在某种程度上减轻人工劳动强度，提高了生产率。

After cocoon cooking comes the silk reeling process, in which the silk is extracted from the cocoon. The purpose is to: ①find the thread of the boiled cocoon through yarn-shaking; ② obtain silk strips through the action of silk floss gathering and silk sheath, with several pieces of bave (cocoon floss) combined; and ③ make small rolls of silk pieces through packaging and drying. During the reeling process, attention should be paid to controlling reeling folds. At present, sitting-type reeling machines and automatic reeling machines (Figure 2-12) are widely used, the latter developed on the basis of the former and accounting for a relatively small proportion. The latter' s technological processes are the same as the former' s, reducing labor intensity to some extent and improving productivity.

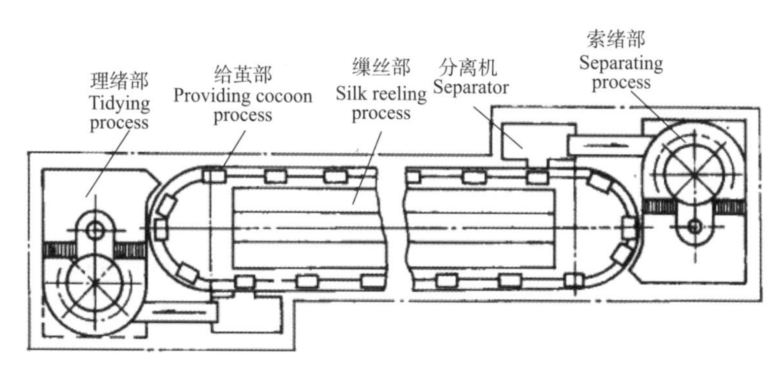

图2-12 自动缫丝机
Automatic Reeling Machine

缫丝主要包括以下几个过程：

The reeling silk mainly includes the following processes:

（1）索绪。将无绪新茧或生产过程中的落绪茧放入高温的索绪锅内，使索绪帚与茧层表面相互摩擦，索出绪丝，索出绪丝的茧子称为绪茧。

Groping end. Raw cocoons or end dropped cocoons in the reeling process are put into a high-temperature groping end basin, so as to rub the surface of the cocoon layer with a groping brush and make ends groped. The cocoons are known as end-groped ones.

（2）理绪。除去有绪茧茧层表面杂乱的绪丝，理出正绪，理出正绪的茧为正绪茧。

Picking end. Incorrect ends on the layer of end groped cocoons should be removed while correct ends picked; the cocoons are called correct end cocoons.

（3）集绪。将若干粒正绪茧的绪丝合并，从接绪装置轴孔引出，穿过集绪器（又称磁眼），集绪器有减少丝条水分、减少颣节和固定丝鞘位置等作用。

Gathering end. Ends of correct end cocoons are combined as a single thread, and the reeling thread passes through (tiny aperture of) end holding apparatus, and end gathering device. The device has the functions of moisture removal of the filament threads, reducing the knots and fixing the position of the filament sheath.

（4）捻鞘。丝条通过集绪器、上鼓轮、下鼓轮后，利用本身前后两段相互捻绞成丝鞘。

Twisted sheath. After the reeling thread passes through the end gathering device, up cam and down cam, the front and back parts of the thread are twisted into the thread sheaths.

（5）缫解。把正绪茧放入温度40℃左右的缫丝汤中，以减少茧丝间的胶着力，使茧丝顺序离解，如离解中的茧丝强力小于其间的胶着力，就会产生断头，这个现象称为落绪，茧子缫至蛹衬而落绪的称为自然落绪，缫至中途而落绪的称为中途落绪。

Unwinding silk. Correct end cocoons are put into a 40 ℃ water bath to reduce the sericin between baves, and the boiled cocoons are dissociated. If the cocoon silk strength in dissociation is less than the sericin between baves, breakage occurs. This phenomenon is called end dropping, including normal dropping and midway dropping.

（6）添绪。当茧子缫完或中途落绪时，为保持生丝的纤度规格和连续缫丝，须将备置的正绪茧的绪丝添上，称为添绪。立缫机用人工添绪，自动缫丝机由机械添绪，由接绪器完成接绪。由于一粒茧的茧丝纤度粗细不一，为保证生丝质量，立缫添绪时除保证定粒外，还必须进行配茧，即每绪保持一定的厚皮茧和薄皮茧的数量比例。

End feeding. The ends of standby correct end cocoons should be fed to maintain the size of the raw silk and continue reeling when the cocoon reeling is finished or dropped in the middle. Sitting-type reeling machines require manual end feeding while automatic reeling machines mechanize the process of end feeding and holding. Cocoon filaments of a cocoon have different sizes. Thus, to secure the quality of raw silk, cocoon baves with different sizes should be prepared beside a given size set up, namely that a certain proportion of thick-skin cocoons and thin-skin cocoons should be maintained for each end.

（7）卷绕和干燥。由丝鞘引出的丝，必须有条不紊地卷绕成一定的形式。丝条通过络交器卷绕在小籰上的称为小籰丝片，其中籰是一种储存丝的装置，卷绕在筒子上的称为筒装生丝。但是无论何种卷绕形式，在卷绕时都要进行干燥。在自动缫丝机上，除了上述各项程序外，新茧的补给、给茧、纤度感知、添绪以及落绪茧的收集、输送和分离等，都由机械来完成。

Winding and drying. The thread drawn from the sheath must be wound in an orderly manner into a certain form. The silk threads are wound by a traversing device onto a reel, and this is called raw silk in a small reel; winding threads onto a bobbin is known as raw silk in a bobbin. However, any form of winding adopted, the threads should be dried when wound. The automatic reeling machine mechanizes cocoon supplying, cocoon feeding, size detecting, end feeding; and gathering, transmitting, and separation of the end dropped cocoons.

五、复摇与整理/Rereeling and Finishing

复摇的目的是把小籰丝片返成大籰丝片，且丝片整形良好，处理成适干程度，保持生丝具有良好的强伸力和抱合力等物理力学性能。复摇过程需要把丝片制作成符合标准的长度和宽度，大绞丝宽70～75毫米，小绞丝宽65～70毫米，长绞丝宽75～80毫米，另外干燥程度需要适当，大籰丝片要求回潮率8%～9%，经温湿度平衡后达到10%～11%，丝片要求络交花纹整齐，同时剔除缫丝中造成的大部分疵点。

Rereeling is a process of reeling the raw silk from small reels onto large reels with good finishing and proper dryness, and keeping raw silk with good mechanical properties such as tensility and cohesion. This is a process where the reels are made according to the width and weight that meet standards. The width of the big reel is 70–75 mm, small 65–70 mm, and long 75–80 mm. In addition, the drying degree should be appropriate. The moisture regain rate of the large reels is required to be 8% to 9%, and after temperature and humidity balance, they can reach 10% to 11%. The reels need to be neatly crossed with the pattern, and most of the defects caused in the silk reeling are eliminated.

整理工序包括编检、大籰丝片平衡、绞丝、称丝、成包入库。编检和平衡主要是检查丝片质量并进行平衡调湿。绞丝和称丝是将平衡后的丝片按照绞装形式绞好并进行称重，一般要求大绞丝125克，小绞丝66.7克，长绞丝180克，称重是为之后计算缫丝工产量提供数据，另外，整理之后的绞丝可以储存用于之后的织造过程。

The finishing process comprises the following steps: weaving and inspecting, balancing the big reels, twisting, weighing, and packing the reels into a warehouse. The main purpose of detecting and balancing is to check the quality of reels and adjust the humidity in balance. The balanced silk reels are twisted and weighed according to the stranded form, which generally requires 125 grams of large reels, 66.7 small, and 180 long. The weighing is to provide data for calculating the output of reeling workers and reeling waste. In addition, the finished twisted silk can be stored for later weaving.

◯ 第三章 / 丝绸织造技艺
Silk Weaving Techniques

第一节 丝织准备与织造工艺 /Preparation and Process of Silk Weaving

一、丝织准备工程 /Silk Weaving Preparation

丝绸的使用价值和审美价值主要由织造工序来实现，可以说，中国丝绸之所以能享誉世界几千年而且经久不衰，是与高超的丝织技艺分不开的。一般而言，丝织可分为两部分：一是丝织准备，二是织造工艺。各类丝织工艺流程的主要区别在于织前准备过程，而织造和检验后入库的织坯整理过程大体相似。在织前准备中，原料要按标准进行检验，以便合理使用。织前准备过程随原料和织物的特点而不同。

The utility and aesthetic value of silk are mainly realized in the weaving process. Chinese silk has been globally well-known for its superb silk weaving-skills for thousands of years. Generally, weaving silk can be divided into two parts: silk preparation and weaving process. The main difference between various silk weaving processes lies in the preparation process before weaving, with the weaving process bearing resemblance to the finishing process of gray fabric in the warehouse after inspection. In the weaving preparation process, the raw material shall be subject to silk quality inspection according to established standards. The process varies with the characteristics of the raw materials and the fabrics.

织物是由经纱和纬纱相互垂直的两组纱线在织机上交织而成的，对于丝织物或其他种类纤维为来说均一致。丝织造的准备工程就是分别组合经纬两组丝线，将经丝按工艺规定的幅宽、长度和根数卷绕成织轴，将纬丝加工成纡子，以满足丝织需要。丝织准备工程一般来说包括：络丝前准备、络丝、并丝、捻丝、定形、整经、浆丝、穿经和结经、卷纬。

The fabric is formed by interweaving two groups of warp and weft mutually perpendicular yarns on a loom, a feature applicable to silk fabrics or other types of fibers. The preparation of silk weaving is to combine warp and weft silk filaments respectively, with warp ones wound into weaving reels according to the width, length and number specified by the process, and weft ones processed into pirns to meet the requirements of silk weaving. Generally, the preparation includes preparation before silk winding, silk winding, doubling, silk throwing, twist setting, warping, warp sizing, drawing–in, warp tying and weft winding.

1. 络丝前准备 /Preparation before Silk Winding

在织造过程中，经丝要求抱合性能好，洁净度高，纬丝要求手感柔软，条干偏差小。因此络丝前的准备工作需要通过浸渍均匀软化丝胶，浸渍是蚕丝加工的独特工艺，能使丝胶软化，丝身柔软光滑。随后进行脱水、蓬松和干燥，蚕丝是容易吸湿的原料，在加工过程中须保持一定的回潮率，因此需要在主要工艺过程之间插入干燥过程。

In the weaving process, warp yarn requires good cohesion and high cleanliness, and weft yarn requires a soft handle and small size deviation. Therefore, the preparation before winding silk needs to soften sericin by dipping evenly. This is impregnation, a unique silk processing technique that can soften sericin and make silk' s body soft and smooth. It is followed by dewatering, fluffing and drying. Silk, a sort of raw material that is easy to absorb moisture, must maintain a certain degree of moisture regain during processing. Therefore, the keeping–dry technique needs to be inserted between major processes.

2. 络丝 /Silk Winding

络丝是把绞装、饼装、筒装的丝卷绕成筒子丝。

Silk winding is the process of winding the silk filaments packed in different forms onto a bobbin.

3. 并丝 /Doubling

并丝类似于棉纺中的并条，主要目的是降低丝条干不匀率，可以分为有捻并和无捻并。

Doubling, similar in cotton spinning to the tie, aims mainly to reduce the unevenness of the filament yarns, which can be divided into twisted and untwisted.

4. 捻丝 /Silk Throwing

捻丝则是为了提高丝条的抗折能力、张力和耐磨性能，使穿着具有凉爽感。并丝和捻丝在丝织生产中占有重要的地位，常可以用来改善产品性能和外观。

Silk throwing is to improve the fracture resistance, tension and abrasion resistance of evenness, making it cool to wear. Doubling and silk throwing are very important in the production of silk weaving, which can often be used to improve the performance and appearance of silk products.

5. 定形 /Twist Setting

定形是加捻后的必要工序，通过加热、给湿等方法加速纤维的弛缓过程，消除内应力，使捻度稳定。

Twist setting is a necessary procedure after silk throwing. The fibers prepared for the next procedure are well processed by heating, wetting and others, with fiber stress eliminated and twist stabilized.

6. 整经 /Warping

整经过程与棉纺的整经过程类似，将卷绕在筒子上的丝线，按总经数、门幅、长度等织物规格平行卷绕成织轴，以便于经丝上浆。

The warping process is similar in cotton spinning to the warping process, where the silk filaments on a bobbin are wound into weaver' s reels in parallel according to such fabric specifications as a total number of warp yarns, fabric width, and length, a process facilitating the weaving of warp sizing.

7. 浆丝 /Warp Sizing

在丝织过程之前，还需要经过浆丝过程，以抵御反复的拉伸摩擦。

Before the silk weaving process, it is also necessary to go through the sizing process to resist repeated tensile friction.

8. 穿经和结经 /Drawing–in and Warp Tying

穿经和结经是经丝准备工序中的最后一道工序。穿经是在翻改织物品种时，按照织物组织将经丝穿入综框的综眼和钢筘内，若有停经片，还需要穿入停经片。结经是不更改织物品种时需要，将满卷织轴的丝头和上机织轴的丝头依次打结，不必再穿经。

Drawing–in and warp tying is the final process. The former refers to the process in which warp yarns are pierced into the hole of the heald frame and reed according to the fabric weave when changing the fabric variety; they need piercing into that of droppers if available. The latter is a process where the silk filament head and tail are knotted in turn when the fabric variety remains unchanged, and therefore there is no need for drawing–in.

9. 卷纬 /Weft Winding

卷纬（俗称摇纡）是把筒装、饼装丝卷成适合织造要求的梭子形状的纡管，它决定了织物纬丝的质量等。

Finally, weft winding is to wind the silk yarns into a shuttle–shaped pirn suitable for weaving requirements, which determines the quality of weft yarns.

二、织造过程 /Weaving Process

无论是现代或是传统织造过程，织物的形成均依赖于开口、引纬和打纬三种机构运动的相互交替，并通过送经和卷取完成最后卷装。

The formation of the fabric in the weaving process, modern or traditional, depends on the alternation of the three mechanisms of shedding, shed weft insertion and beating–up, and the final package is completed through let–off and takeup.

开口运动是根据织物组织要求带动经丝上升或下降，将织口和停经架间的经丝上下分开形成梭口，便于之后将纬丝导入，常见的机构包括踏盘开口机构、多臂开口机构、提花机开口机构。

Shedding is to drive warp to ascend or descend according to the requirements of fabric texture, with the top and bottom of the warp yarn between cloth fell and warp stop motion separated to form the shed for the later introduction of the weft yarn, with tappet, dobby and jacquard motions available.

引纬运动是纬丝在水平方向的运动，对于有梭织机来说，是将装有纡子（纬丝）的梭子穿入梭口，除此之外，还有其他的引纬方法，包括剑杆织机、片梭织机、喷气织机和喷水织机等，采用除梭子外的其他方式引入纬丝。

Weft insertion is in the horizontal direction of the weft. In terms of weft insertion motion such as a shuttle loom, a shuttle with pirn is thrown into the shed. However, other weft insertion motions, including rapier loom, projectile loom, air–jet loom and water–jet loom, adopt different ways to introduce weft yarns.

打纬运动是将导入梭口的纬丝打向织口，在织口处使经、纬丝紧密交织形成织物。常见的机构包括四连杆打纬机构和共轭凸轮打纬机构。

Beating–up is to drive the weft introduced into the shed toward the cloth fell, with the warp and weft filaments closely interwoven at the cloth fell to form fabric. The beating–up mechanism includes a four–bar beating–up one and a conjugate cam beating–up one.

送经运动是使织轴转动送出经丝，补充形成织物所消耗的经丝长度。

Let–off is to make the weaver’ s beam rotate so that the warp yarns can be sent off, supplementing the length of the warp consumed to form a fabric.

卷取运动是将形成的织物引离织口，并将其卷绕在卷轴辊上，以保证制造工序的连续进行，有两种卷绕机构，分别为消极式卷曲机构和积极式卷曲机构。

Take–up involves leading the formed fabric away from the cloth fell and winding it on the reel, to ensure that the manufacturing process is continuous, with two winding mechanisms available, negative and positive.

这五大运动都是在织机主轴一回转时间内完成的，并周而复始地重复进行。除此之外，织机还有多梭箱机构，自动引纬机构，防织疵和安全机构，启动、制动机构以及传动机构。在织物形成过程中，当钢筘将纬丝打向织口时，产生经纬丝之间的相互摩擦和弯曲，经丝反复经受拉伸和弯曲，而后梁、综眼及筘齿处还受到摩擦力的作用。

These are all completed within the rotation of the loom spindle and repeated over and over again. In addition, there are multi-shuttle mechanisms, automatic weft insertion mechanisms, anti-blemish and safety mechanisms, starting and braking mechanisms, and driving mechanisms. When the reed hits the weft towards the cloth fell during fabric formation, friction and bending between the warp and weft filaments are generated, with the warp repeatedly stretched and bent and friction force applied to the rear beam, heald eye and dent.

市面上丝织品种类繁多，主要是由于织造技术以及织物结构不同导致的。目前，丝织物中提花织物和多色织物的比重较大，所以提花开口机构和多梭箱织机的应用比较普遍，例如，中国像景丝织物是在特制的多梭箱像景丝织机上织制的。20世纪50年代已经用自动割绒织机织制起绒丝织物，如乔其绒、金丝绒等。有些具有手工艺品特色的产品仍须用手织机织制，如雕花天鹅绒，起绒部位要使用起绒杆，下机绸坯经手工雕花割绒。中国妆花织品是用多种彩色纡子挖花织制的，能在同一纬向花纹中显示多种纬色。幅度较窄的腰带织物也是在手织机上按设计纹样用小梭子挖花织制而成，成品通经断纬，花纹复杂，色彩丰富，缂丝也是这样挖花织制的。

The differences in weaving technology and textile structure are the cause of a wide range of silk fabrics available in the market. At present, a relatively large proportion of jacquard fabric and multi-color fabric boosts the wider application of jacquard shedding mechanism and box loom. For example, Chinese photographic fabric is woven on a special box loom. Fleece fabric, such as georgette velvet and pleuche, was woven by automatic velvet cutting loom in the 1950s. Some products can merely be woven by hand, such as carved velvet, with its gigging part put into raised rod, and the greige carved and cut by hand. Chinese Zhuang Hua silk fabrics that are woven in a variety of colorful pirns with cutwork techniques display a variety of weft colors in the same weft direction pattern. The belt fabric with a narrow width is also woven on a hand-woven machine according to the pattern design with small shuttle cutwork, with its pattern complex and color-rich. The same is true with kossu.

在未来的丝织生产中，增加花色品种，提高质量，发展高速高效、大卷装和连续化的设备，如采用并捻联合机、整浆联合机简化工艺过程，是重要的发展方向。若采用复动式全开梭口和连续纹板的提花织机，配以弹簧回综装置，能大幅提高织机速度。

In the production of silk weaving, attention should be paid to the following aspects: the variety of patterns and colors; the quality; and the equipment for process simplification with high speed, high efficiency, large capacity and continuation, such as doubling and twisting machine, and warping-sizing machine. The acceleration of the looms can be achieved with the jacquard loom with double lift full-opening shed and continuous pattern plate, equipped with a spring weight device.

第二节　传统丝织机 / Traditional Silk Weaving Machines

中国丝织工艺以其历史悠久、技术先进、制作精美著称于世。我国早期的纺织技艺是从编筐席或编发辫中得到启发。将准备好的丝线中的经线依次绑在两根平行且保持一定距离的木棍上拉紧，然后将纬线依次类似编席那样横向织入一根根经线中，这样经、纬线纵横交错便织成织物。这种原始的织造方法在古代叫作“手经指挂”，意思就是徒手编织，没有借用任何织具，完全是由人的手或指直接操作完成的。据传说，这种“手经指挂”的编织技艺产生于1万年前的旧石器时代，从河姆渡遗址和西安半坡遗址等的编席以及有编织物印痕的陶片中可以证明，这种“手经指挂”的织造方法一直沿用到新石器时代。

Chinese silk weaving is well known for its history, advanced technology and exquisite manufacturing. The earliest textile skills were inspired by weaving basket seats or braids. The warps in the prepared silk threads are tied on two parallel wooden sticks with a certain distance in turn to be tightened, and then the wefts are woven into the warps in a transverse direction like mat weaving in turn so that the warps and the wefts can be interlaced vertically and horizontally to be woven into silk fabric. This primitive weaving method entitled Shoujing Zhigua (warps hanging on the fingers for weaving) means weaving with bare hands. Legend has it that this primitive technique was born 10,000 years ago in the Paleolithic Age, a technique used until Neolithic Period, as demonstrated by mat weaving and ceramic chips with knitting marks preserved in the Hemudu site and Xi’ an Banpo site.

我国古代的文献资料中，记载的丝织品种类繁多，如锦、缎、绸、绢、绫、罗、纱等，而且文献中对这些丝织品的特点和风格都有所说明，其中一些织物及其名称一直沿用至今，此时的丝织物上还有菱形纹、回纹和畦纹等几何纹图案。这充分说明，我们的祖先在商周时期已经意识到织物的组织结构决定着织物的花色品种和外观风格，并能熟练地通过细度、密度、捻度、捻向等参数的变化使平纹织物产生相应的变化，由此创造开发了众多的织物品种。与此同时，还在平纹组织的基础上创造了斜纹组织，宋朝又出现了缎纹组织，此时，三原组织（平纹、斜纹、缎纹）都已出现，为我国古代织物组织的发展变化奠定了基础，这在世界纺织史上是一个很大的进步。

There are many kinds of silk textiles recorded in ancient China, such as brocade, satin, silk, luster, silk twill, half-cross leno and yarn, the characteristics and styles of which are discoursed in the literature, with some of the textiles and their names kept to this day. The silk textiles in ancient China boast various geometric patterns such as diamond, fringe and twisted polka, an indication that the ancestors in the Shang and Zhou Dynasties were aware of the effect of the structure on fabric variety and appearance, who could skillfully change the plain weave fabric through changes in parameters such as fiber fineness, density, twist, and twist direction, thereby creating a large number of fabric varieties. In addition, twill weave was created on the basis of plain weave, with satin weave appearing in the Song Dynasty, a time when the three fundamental weaves (plain, twill and satin) had appeared. This laid the foundation for the development of textiles in ancient China, great progress in the history of world textiles.

一、腰机/Waist Machine

在新石器时代，我国的纺织技术有了很大的发展，在手工编织的基础上发明了原始的织布机及相应的织造工艺。原始的机织技术是用骨针进行手工编织，但速度很慢，达不到所需的紧密程度，经过长期的实践，原始腰机的织造过程逐步发展起来（图3-1）。

In the Neolithic Period, Chinese textile techniques had been greatly developed, with the primitive loom and corresponding weaving process invented on the basis of hand-knitting skills. The primitive weave was hand-knitted with bone needles, but the speed is too slow to reach the required tightness. After long-term practice, the weaving process of the primitive waist machine was gradually developed (Figure 3-1).

图3-1 腰机织造示意图
Schematic Diagram of Weaving Process of Waist Machine

1975年，浙江余姚河姆渡遗址出土了几件木制织造工具，包括卷布辊、打纬刀、分经杆和提综杆这几件木制器具，这些都和腰机上的织造工具相一致，说明在新石器时代我国就已经拥有了腰机织造技术。这些工具特别之处就是在纺织技术史上创造性地使用了综杆和分经装置，把经线简单地分为上下两层，一次引纬动作就可把纬线穿过所有经线形成的梭口，这就将生产效率大大提高。当时的腰机已拥有开口、引纬、打纬、送经和卷取五大功能，它的使用是需要织工席地而坐，将卷布辊系在腰间，利用腰背和双脚前挺后摆进行织造，一般用于制作平纹织物，也具有提花的功能，目前此类腰机在我国少数民族地区还有保存。腰机织物与现代生产的电力纺产品十分接近，说明了我国织机技术的先进性。

In 1975, several wooden weaving tools were unearthed at Hemudu site in Yuyao, Zhejiang Province, including a cloth roller, beating-up knife warp bar and harness lever. These are all consistent with the weaving tools on the waist machine, evidence that China already developed the waist weaving technique in the Neolithic Period, particularly the creative use of harness lever and warp dividing devices in the history of textile technology. With these tools being applied, warp yarns are simply divided into upper and lower layers, by which the weft yarns pass through the shed formed by the warp yarns with only one weft insertion. Accordingly, the textile production efficiency rises greatly. The waist machine then had five functions shedding, weft insertion, beating-up, let-off and take-up. Running this machine requires that the weaver should sit on the ground, tie a cloth roller around the waist, and make better use of the waist, back and feet to weave. It is generally used to make plain weave and jacquard. Such machines are still preserved in areas of ethnic minorities in China. Waist machine fabric is very close to spinning products by modern machines, suggesting that Chinese textile technology was advanced.

二、罗织机/Half-cross Leno Loom

河南青台出土的新石器时代后期罗纹丝织物表明，我国早在6000年前就有了织造罗纹的技艺。罗纹织物历史悠久，商周时期以二经相绞的素罗为主，之后逐渐发展出链式绞组织，绞经轮流用左侧或右侧的地经交替，环环相扣，呈现不可分割结构。唐宋时期，素罗和花罗用途广泛，生产量较大，但在元末明初链式罗逐渐失传。古代链式罗结构不同于一般纱罗织物，其外观与针织品类似，其厚度和透明度介于纱和绸之间，光泽、透气性均佳。

The late Neolithic Period rib-stitch silk fabric unearthed in Qingtai, Henan shows that China had the skill of weaving leno as far back as 6,000 years ago. The leno fabric has a long history. There was mainly plain leno in the Shang and Zhou Dynasties, a leno woven in the way that two warp yarns (crossing warp and ground warp), are twisted regularly. There gradually developed a chainlike leno in which the crossing warp is interwoven with the ground warp by alternating with the left or right side, forming a tight structure. Plain leno and patterned leno were widely used and in large quantities in the Tang and Song Dynasties. The chainlike leno, however, was lost at the end of the Yuan Dynasty and the beginning of the Ming Dynasty. Different from general leno fabric, ancient chainlike leno has a unique structure, good gloss and air permeability, with its appearance similar to that of knitwear, and thickness and transparency lying between yarn and silk.

关于罗织机的结构，目前只能通过古书记载得知，薛景石的《梓人遗制》展示了罗织机的图形，如图3-2所示。该罗织机具有一套简易的送经和卷取装置，较为特殊的是开口机构，采用半综装置，一端系吊综绳连脚踏杆，另一端鸟眼下吊绞综范子（架）。若织造提花罗，须增加提花机构（束综）来控制花型。另外，斫刀是织造罗织物必不可少的工具，由于罗织物经线相互缠绕，无法用普通织机上的筘来织造，斫刀可以完成引纬和打纬的工作。当梭口较低且不清晰时，插入纹杆可起到清晰梭口的作用，因此纹杆是顺利织造罗织物的关键。

The structure of the half-cross leno loom can only be found in ancient books, such as *Ziren Yizhi*, known as *Practice of Carpentry* written by Xue Jingshi, presenting the graphic of the loom, as shown in Figure 3-2. The half-cross leno loom has a set of simple let-off and take-up devices, with a shedding mechanism specially designed, namely that the half-heald (doup) device is adopted, with one end of the wooden bar, a bird sitting on a piece of wood, tied with the pedal bar by harness mounting, and the other end linked with the doup harness frame. If the jacquard half-cross leno is weaved, the jacquard mechanism must be added to control the pattern of the flowers. In addition, the cutter, an indispensable tool for weaving half-cross leno fabric, functions well regarding shed weft insertion and beating up. Because the warp yarns are wound with each other, reed on an ordinary loom cannot be used for weaving. When the shed is low and not clear, rod insertion is useful for clearing the shed. The rod is thus the key to weaving the fabric.

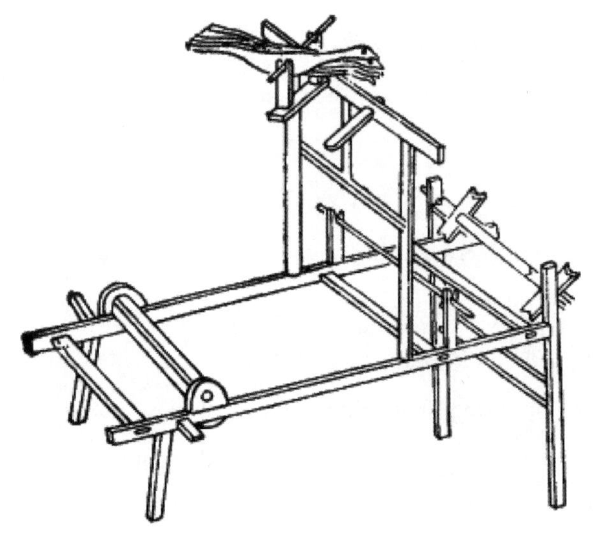

图3-2 《梓人遗制》中的罗织机
Half-cross Leno Loom in *Ziren Yizhi*

三、斜织机/Twill Loom

斜织机（图3-3）是用于制作平纹织物的织机，其使用、流传时间久远。它是在原始织机的基础上演变而来，并逐渐改进与完善，织物规格趋向统一，纺织生产也逐步走向职业化，如用打纬刀代替引纬，并发明了筘，用来控制经面的宽窄和边撑，对布幅进行有效的控制，随后就逐渐地代替了打纬刀，后来又被梭子所代替。斜织机起源于何时目前仍无法得知，春秋时期的文献记载了一种鲁机的织机结构，它就是在原始腰机的基础上增加了机架、定幅筘、经轴，整体更为完整，但鲁机是否属于斜织机并没有统一的结论。

Twill loom (Figure 3-3) has long been used for making plain fabrics. Developed on the basis of the primitive weaving machine, it has been improved, with the specifications more unified, and production gradually moving towards professionalization, such as the substitution of weft insertion with weft beater, the invention of reed for controlling the width and the edge brace of warp flush, and effective regulation of the cloth width. The reed was later replaced by a shuttle. It is still little known as to the origin of the twill loom. However, the structure of the Luji loom was recorded in the literature in the Spring and Autumn Periods. Added to the primitive waist machine are a frame, spacing reed and warp beam, a more complete mechanism. However, it remains uncertain whether Luji is twill loom.

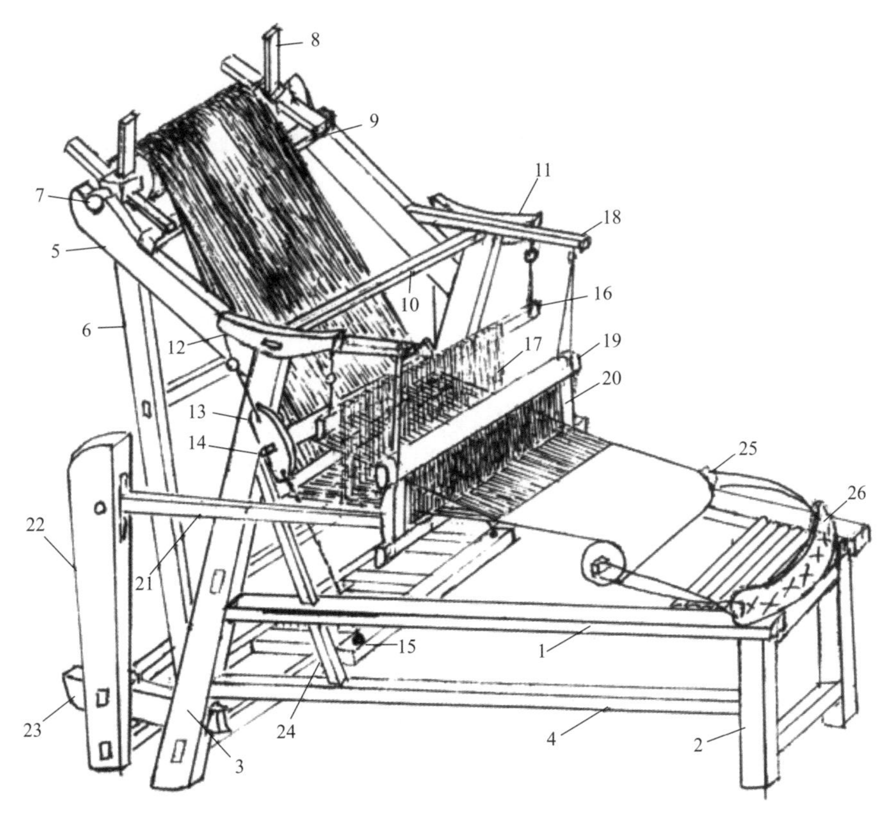

图3-3 斜织机示意图
Schematic Diagram of Twill Loom

1—上排沿/Up edge row 2—前机腿/Front stand leg 3—后机柱/Rear engine column 4—下排沿/Down row edge 5—滕子架/The shelf of wrap yarn 6—支翘撑/Support frame 7—滕轴蕊/The central spindle of wrap yarns 8—羊角桩/Cleat pile 9—搭角枋/The square-column supporting cleat pile 10—摆杆梁/Pendulum beam 11—摆动臂/Swinging arm 12—牵导索/Pulling cable 13—导辊片/Roll guide 14—压经辊/The roll pressing wrap yarns 15—综蹑板/The pedal controlling heddles 16—提综杆/The rod pulling heddles 17—线综环/Heddles 18—吊框架/Hanging frame 19—筘框帽/Reed cap 20—筘闸板/Reed flashboard 21—牵手臂/Dragging arm 22—叠木座/Pedestal 23—叠助木/Assistant wood 24—斜支撑/Diagonal support 25—卷布辊/Cloth roller 26—腰皮带/Belt

斜织机的操作者不再需要席地而坐，因为机座离地有一定距离，且机前机后距离不同，略向前倾斜，机座面与经丝面成50°～60°，斜向设计更有利于发现经纱疵点。另外，根据当时史料记载，丝织品生产率大幅度提高，传统腰机难以胜任。斜织机设置了持经机构、开口机构和打纬机构，且更加自动化，可通过脚踏方式提综，同时保留原有腰机的卷布辊，因此引纬、送经和卷取操作仍需手工完成。

Operating a twill loom, the weavers do not sit on the ground, because the frame is at a certain distance from the ground. The loom tilts forward slightly, with the frame surface and the warped surface forming a 50–to–60 degree. Such design is more conducive to the inspection of the fabric surface defects. In addition, according to historical records at that time, the productivity of silk fabrics was greatly improved, beyond the reach of traditional waist machines. The twill loom, equipped with a warp holding mechanism, a shedding mechanism and a beating–up mechanism, is more automatic so that the heald lifting can be done by pedaling. However, the functions of weft insertion, let–off and take–up still need to be completed manually because the cloth roller of the primitive waist machine is retained.

四、多综多蹑提花织机/Multi–heddle and Multi–treadle Loom

早在殷商时期，我国就能织造复杂几何图案的丝织品，而这样的丝织品并非单综或者双综的制作平纹织物的斜织机能够完成的，制作者需要掌握手工腰机及多综提花技术才能织造。战国时期，出现了脚踏提综技术，利用多根脚踏杆控制综片的上下运动，因此，产生了多综多蹑提花织机，此时的脚踏织机已达到非常完善的程度。在春秋战国时期，我国织物提花技术得到很大的发展，出现了许多复杂多变的鸟、兽、龙、凤花纹，而这些花纹都需要很高的织造技艺和更为复杂的织机完成，这就充分说明当时的织机和织造技术有了很大进步。到了汉朝，丝织物的花纹变化更加繁多，综片和蹑的数量大大增多，机型也日趋完善。先前为了织制高档贡品，综数高达150综，但工艺复杂且效率过低，因此后期降到50综左右，产量大大提高，具有广泛的应用前景。多综多蹑织机占地较小，技术易掌握，所以被长期使用，与束综提花织机并存，现今在我国西南少数民族地区仍用来织造花边和腰带等。

As early as the Yin and Shang Dynasties, silk fabrics with complex geometric patterns could be woven by both manual waist loom and multi–harness jacquard loom instead of twill loom, one that is merely applicable to making plain fabrics with single or double harnesses. During the Warring States Period, the technique of heddle lifting by pedal gradually appeared, by which multiple pedal bars are used to control the up–and–down motion of the heddle. Accordingly, multi–heddle and multi–treadle loom came into being, with the pedal loom well developed. During the Spring and Autumn Period and the Warring States Period, the jacquard technology made unprecedented progress, with many complex and various patterns available, such as birds, wild animals, dragons and phoenixes. These patterns were skillfully woven with more complex weaving machines, suggesting that the machines and techniques grew more advanced at the time. The patterns increasingly changed in the Han Dynasty, with the number of heddles and peddles greatly rising and models being increasingly optimized. Previously, weaving high–end tributes was complex, with the number of heddles as high as 150, resulting in low efficiency. Therefore, the number of heddles in the later period dropped to about 50, and the output was greatly improved. This optimized technique was suitable for wide application. Multi–heald and multi–peddle loom has been used for a long because of its smaller size and user–friendly operation. As a result, the loom, together with the jacquard loom, is still used to weave lace and belt in the southwest region of ethnic minorities.

五、束综提花织机/Jacquard Loom

在东汉时期，我国就已出现了带花楼的织机，也就是前段提到的束综提花织机，之所以称之为花楼织机，是因为这种提花织机比一般综片机高出一个束综提花装置，织造者从下向上看似高楼双峙，若坐在离地三米多高的花楼上，千丝万缕的光亮经面犹如下临清池，因此花楼织机是按照其形态命名的。

A loom more than three meters in height, jacquard loom already existed in the Eastern Han Dynasty, one with an additional jacquard device that bears resemblance to a special double–tower building if seen from a weaver looking up.

在多综多蹑织机发展的过程中，受到综片数量的影响，纬循环数也不能过多，因此花纹图案不能过大，后来就出现了束综提花织机，开口部分不再使用综片，而是将每根经线用线综牵吊。织造者按照花本编制的程序，在花楼上进行牵拉提沉完成开口运动，这样纬循环数可大大增加，花纹可以扩大，图案样式更加丰富。花楼可分为大花楼和小花楼，视花纹复杂程度而定，同时花楼织机也具有素综装置，一位织造者用脚踏机构控制素综，另一位织造者按照花本程序来提花。

During the development of multi–heald and multi–peddle loom, the number of weft repeat unit is limited by the number of healds, so the pattern should not be too large. Later, the jacquard loom appeared, and healds are no longer used at the opening part. Instead, each warp thread is pulled and hoisted with a heald. According to the pattern draft, the weavers finish the shedding motion with the jacquard loom. By doing so, the number of weft repeat unit greatly increases, with the pattern expanded and abundant. The jacquard can be divided into a large jacquard loom and a small jacquard loom, depending on the complexity of the patterns. In the jacquard process, the decoration system and the knitted jacquard are mutually compatible and both are indispensable.

Two weavers work together: one sits on jacquard to weave the jacquard according to the pattern draft, and the other steps on the peddle to weave.

三国时期，发明家马钧对束综提花机又做了许多改进和简化，再经过唐、宋几代的改进提高，逐渐完善而定型。唐朝以前虽有束综提花织机的记载，但并没有图像留存，宋元时期的《蚕织图》上有关于花楼织机的描述，该设备为花楼束综绫织机，如图3-4所示。明清之后直到近代，这种花楼织机的织造工艺一直在各地长期保存和使用。中国古代的提花机直至明代一直处于世界领先地位，其高超的挑花结本原理为近现代的穿孔纹板和计算机控制提花提供了借鉴，这说明我们的祖先为人类进步做出了巨大贡献。

During the Three Kingdoms period, inventor Ma Jun made many improvements and simplifications to the jacquard machine. The machine was gradually perfected and finalized after the Tang and Song Dynasties. Although there were records of the jacquard loom before the Tang Dynasty, there are no images left. A jacquard loom depicted on a silk weaving book entitled *Silk Weave Diagram* in the Song and Yuan Dynasties, as shown in Figure 3-4, is another type of jacquard loom, silk twill loom, characterized by thread-based flowers used for weaving silk twill. From the Ming and Qing Dynasties to modern times, the weaving process of this jacquard loom has been maintained and used for a long time in various places. The jacquard machine in ancient China had been in the leading position in the world until the Ming Dynasty. Its superb principle above provides a reference for modern electronic jacquard. There' s no denying that our ancestors made great contributions to mankind.

图3-4 《蚕织图》上的花楼束综绫织机
Jacquard Loom from the Book *Silk Weave Diagram*

六、漳缎织机 /Zhangchow Velvet Satin Loom

除之前所述的普通大花楼和小花楼织机外，还有一种漳缎织机（花绒织机），它是我国古代花楼织机中机械功能较为完善、技术工艺较成熟的大花楼机，可用于编织各种各样的提花。与其他提花织机不同的是，漳缎织机采用了独立式挂经装置，解决了普通花楼织机无法纺制花绒织物的技术难题，在显花和起绒方面具有独特性。漳缎织机的起绒织造工艺技术至今仍被沿用。漳缎织机分为旱机和坑机两种，在我国南方地区，降雨量大，地势低洼，坑机易积水影响作业，一般采用旱机来织造，但坑机整体要优于旱机，因为坑机用料要少于旱机，铺设后整机稳定性强，不易移位，且操作方便，地下湿度也对织造有利。

In addition to the ordinary large and small jacquard looms mentioned before, there is also a Zhangchow velvet satin loom. With the most perfect mechanical functions and the most mature technology in ancient China, it could be used to weave multi–variety jacquard. Different from other jacquard looms, Zhangchow loom adopts an independent warp hanging device, which solves the technical problem that ordinary jacquard loom can’ t spin flower velvet fabric. Therefore, it has uniqueness in flower display and gigging. The gigging weaving process technology of Zhangchow loom is still used today. The loom is divided into two types: dry jacquard loom and pit jacquard loom. In southern China, a lot of rainfall and low terrain is a challenge for operating pit jacquard loom. Generally, dry jacquard loom is used for weaving. However, the pit jacquard loom performs better than the dry jacquard loom as a whole, such as the strong stability of the whole loom, the convenience of operation and ground humidity, which is also beneficial to weaving.

漳缎织机由于拥有独立式挂经装置，整机体积很大，全长达6米，机宽1.2米，从地面至花楼顶端高达3米以上。织机同样拥有开口、引纬、打纬、送经和卷取装置，不过其中的引纬过程需要手工完成。宋朝开始出现的缎纹组织漳缎是我国古代绒类织物的代表作，始于明末清初福建漳州，由两组经线和四组纬线交织而成，原有素绒织物上进行了织物结构创新，成为具有艺术特色的以缎纹为地、绒经起花结构的全真丝提花绒织物。在工艺技术上极为精湛，制作漳缎使用的提花绒织机，是中国古代花楼机中机械功能较为完善、机构较为合理、技术工艺较为成熟的一种，并一直传承。漳缎绒有花漳绒和素漳绒两类。素漳绒表面全部为绒圈，而花漳绒则将部分绒圈按花纹割断成绒毛，使之与未割的绒圈相间构成花纹。

Because of its independent warp hanging device, the Zhangchow velvet satin loom has a large volume, with a total length of 6 meters, a width of 1.2 meters, and a height of above 3 meters. The loom has the same shedding, shed weft insertion, beating–up, let–off, and take–up devices. One of the differences is that the shed weft insertion process needs to be completed manually. Zhangchow velvet with satin weave, which began to appear in the Song Dynasty and dated back to Zhangzhou, Fujian Province at the end of the Ming Dynasty and the beginning of the Qing Dynasty, was a representative work of ancient Chinese velvet fabrics. It was formed by interweaving two sets of warps and four sets of wefts. The innovated version on the basis of the original plain velvet fabric became the artistic jacquard velvet fabric with satin weave as the ground warps. The jacquard velvet weaving machine used in the manufacture of Zhangchow velvet is most complete in the mechanical function, most reasonable in mechanism and most mature in technical process among ancient Chinese jacquard loom. Therefore, it has been passed on to generations. There are two types: patterned and plain Zhangchow velvet. In terms of the latter, its surface is all woven with a loop pile; However, for patterned Zhangchow velvet cuts, its pattern was formed by alternating loop pile and low loop pile which was cut according to the pattern.

七、踏板立机/Peddle Vertical Loom

除上述几种流传久远且经典的丝织机外，还有几种特殊的织机。唐末五代时期出现了踏板立机，敦煌的壁画和文书中都有所提及，在北宋时期的开化寺壁画上清晰可见当时的踏板立机，如图3-5所示。元朝薛景石著的《梓人遗制》，对罗织机的几种织机进行了规范和改革，使其更为完善，其中记载了非常完整的踏板立机资料。此外，明代的《宫蚕图》中也有一台高大的立机。目前关于踏板立机的资料并不多，只有以上4种，其中薛景石记录的薛式立机非常具有分析价值。

In addition to the above ancient and classic silk looms, there are also several special looms. At the end of the Tang Dynasty and Five Dynasties, there appeared peddle vertical loom, whose warps are perpendicular to the ground, as seen in Dunhuang frescoes and some documents. It can also be clearly seen on the frescoes of Kaihua Temple in the Northern Song Dynasty (Figure 3-5). However, the most detailed record is Xue Jingshi' s *Ziren Yizhi* in the Yuan Dynasty, in which several looms including half-cross leno loom were standardized and reformed. In addition, there is also a peddle vertical shown in *Palace silkworm diagram*. At present, there are only the above four kinds of peddle vertical looms, with few references available, among which Xue Jingshi' s peddle vertical has the most analytical value.

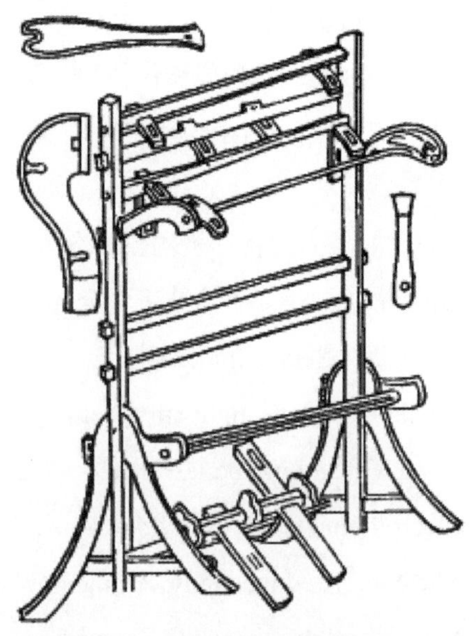

图3-5 开化寺踏板立机
Peddle Vertical Loom in Kaihua Temple

在薛式立机的描述中，每一个零件都有其名，并配有插图进行示意，全机零件共29种。一般织机均可完成之前所述的开口、引纬、打纬、送经和卷取五大运动，而对于薛式立机来说，较为复杂的就是开口机构。薛式立机的脚踏板分为长短两个，长短踏板分别踩下代表着后引手的顶上和拉下，长踏板踏下后，综片被提起，短踏板踏下则综片放松，经丝依靠张力复原。此外，立机的高度可变，一般约为2米。

What matters in Xue' s description is that each of all 29 parts bears its name, together with illustrations. The most complicated design of Xue' s peddle vertical is the shedding mechanism, one of the five major motions described before. The pedals of peddle vertical in Xue' s description are divided into two types: the longer one pedaled, the heddle is lifted; the shorter one pedaled, the heddle is relaxed and the warps recover with tension. In addition, the height of the peddle vertical can be varied, generally about 2 meters.

八、竹笼机/Bamboo Cage Loom

在我国广西壮族自治区，壮锦等提花织物仍采用竹笼机（图3-6）进行织造，竹笼机早在宋朝就已存在，织造工艺也较为独特。竹笼机结构简单，操作方便，它在提综机构上主要以少蹑来控制多综，与多综多蹑有明显区别，在提花机构上，竹笼机采用了竹编笼的形式，由机座底部两根蹑杆来分管竹笼的起落和地综的提升，竹笼上环绕的花纹杆数一般为70～120根，由花纹循环数决定。

In Guangxi ethnic minority area, the jacquard fabric, such as Zhuang brocade (one of the most representative national handicrafts in the Guangxi Zhuang Autonomous Region) is still woven with bamboo cage loom (Figure 3-6), which dates back to as early as the Song Dynasty. The loom, with a unique weaving process, has so simple structure that it is easy to be operated. In terms of the heald lifting mechanism, it mainly works by controlling multiple healds with fewer peddles. In terms of the jacquard mechanism, it adopts a bamboo weaving cage. These two mechanisms were respectively controlled by two peddle bars at the bottom of the loom base. What' s more, the number of bamboo sticks around the bamboo cage is generally about 70-120, which is determined by the number of pattern repeat units.

以织造壮锦为例，左脚蹬花蹑，竹笼将抬起，双手取下竹笼上的竹编纹杆，将单组纹综向后推移，形成梭口，然后根据花色要求进行引纬和打纬。另外，提起竹笼也可以将双组纹综向后推移，使反向花部的经纱提起，形成反向梭口，相当于织制反向花纹。右脚控制地蹑，底层经纱将随环综提起，形成梭口织制平稳地纬。竹笼机的送经和卷取工作也需要人工操作，因此劳动强度较大，需要手、脚和腰背相互配合，动作需协调，产量较低。此外，织物门幅受限，当织造较宽的织物时，需要多幅拼接。

Take weaving Zhuang brocade for example. Step on the left pedal to lift the bamboo cage, take off the weaving rod on the cage and move backward the single group of warp for shedding. Then shed weft insertion and beating-up are required according to the design effect. In negative spinning, the two groups of warps can be moved backward if the bamboo cage is lifted, so that the warp yarns of the pattern can be lifted for negative shedding. The warps of the bottom layer are lifted by a harness lever when stepping on the peddle. This helps to shed to weave ground weft smoothly. It is high labor intensity because the let-off and take-up of operating a bamboo cage loom also require manual work. But the output is low. In addition, when a piece of wider fabric is woven, splicing of multiple pieces is required because of the limited fabric width.

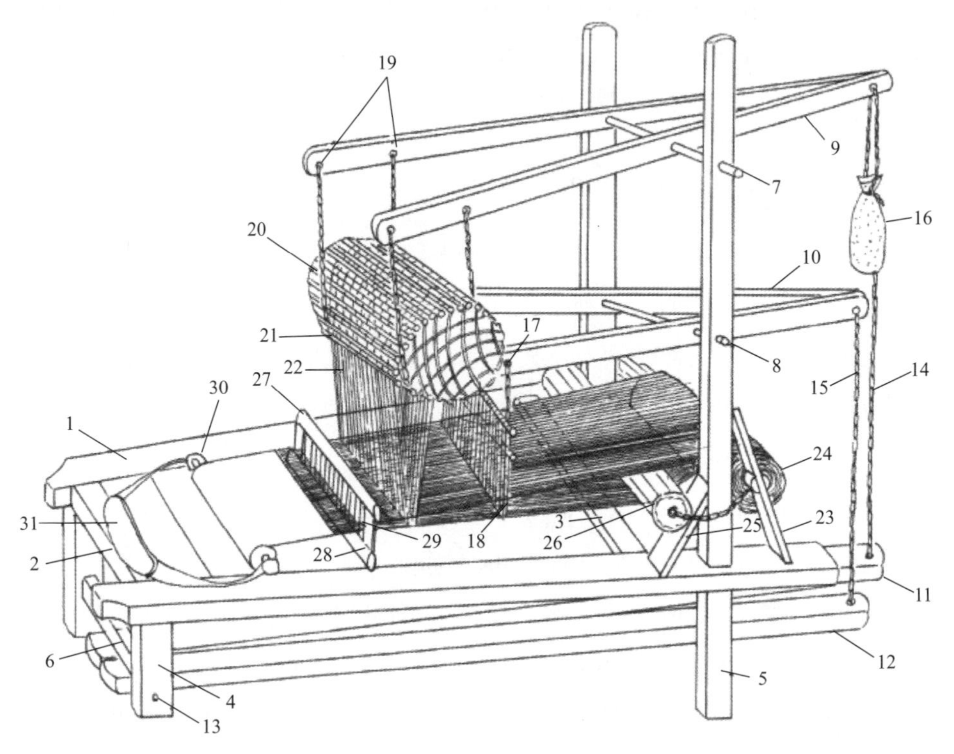

图3-6　竹笼机示意图

Schematic Diagram of Bamboo Cage Loom

1—排雁/Edge row　2—前档/Front stop　3—后档/Rear stop　4—前机脚/Front stand leg　5—联体机柱/Conjoined engine column　6—机脚档/Bottom stop　7，8—摆臂杆肖/The nut of swinging arm lever　9—竹笼摆臂/The swinging arm of bamboo cage　10—地综摆臂/The swinging arm of ground heddles　11—花蹑/The pedal controlling bamboo cage　12—地蹑/The pedal controlling ground heddles　13—蹑杆肖/The nut of pedal rod　14，15—牵引绳/Hauling cable　16—垂袋/Drooping bag　17—地综杆/The rod of ground heddles　18—地综/Ground heddles　19—环笼绳/The rope surrounding cage　20—竹笼/Bamboo cage　21—纹竿/Pattern rod　22—纹综/Heddlesl　23—H形轴板/“H” axillare　24—经轴/Wrap beam　25—斜撑/Diagonal support　26—分经筒/The bin separating wrap yarns　27—筘框/Reed frame　28—竹筘/Bamboo reed　29—框闸栓/The brake bolt of reed　30—卷布辊/Cloth roller　31—腰带/Belt

第三节　典型丝织技艺/Typical Silk Weaving Skills

一、蜀锦织造技艺/Shu Brocade Weaving Skills

蜀锦拥有2000年的历史，多用染色的熟丝线织成，用经线起花，运用彩条起彩或彩条添花，用几何图案组织和纹饰相结合的方法织成，色调鲜艳，对比性强，是一种具有中国汉族特色和地方风格的彩色织锦。四川是桑蚕丝绸业起源最早的地方，是中国丝绸文化的发祥地之一。蜀锦兴于春秋战国时期，兴盛于汉唐，因产于蜀地而得名。在传统丝绸织锦的生产中，历史悠久，影响深远。2006年，蜀锦织造技艺经国务院批准列入第一批国家级非物质文化遗产名录。

Shu brocade with a history of 2,000 years, is dyed boiled-off silk threads, with its patterns woven on the basis of warp. It has a combination of geometric patterns and engraved patterns. This process contributes to bright tones, bringing about strong contrast. So it is a colorful weaving brocade with Chinese Han characteristics and local style. In Sichuan, one of the birthplaces of Chinese silk culture, the sericulture industry originated at the earliest time. Shu brocade rose from the Spring and Autumn Period to Warring States Period and flourished in the Han and Tang Dynasties. Nowadays, it still has a far-reaching influence on the production of traditional silk brocade. In 2006, Shu brocade weaving skills were approved by the State Council to be included in the first batch of the national intangible cultural heritage list.

蜀锦以桑蚕丝为经纬线，按蜀锦生产的工艺过程及规范，经过若干工序的组合，改变桑蚕丝之间的结构形态，使绞装生丝变成了精美细腻、色彩艳丽的蜀锦，这通常被称为蜀锦的传统织造工艺。蜀锦的主要工艺包括图案纹制工艺、经纬丝练染工艺、丝织的准备工艺以及丝织织造工艺四部分。图案纹制工艺是根据织机纹针数、织物经纬密度、纹样类型及纹样大小按物组织绘制意匠图及设色；经纬丝练染工艺，将生丝变为熟丝，即对生丝进行染色；丝织的准备工艺，即经纬丝的卷装及位置安装；丝织织造工艺即采用花楼织机进行织造。

Shu brocade' s warp and weft yarns are made of mulberry silk. They are processed according to the related specification after the process of silk weaving. Then, the skein silk turns out to be exquisite and colorful Shu brocade, which is generally called the traditional weaving process of Shu brocade. The main process of Shu brocade consists of four parts: pattern drafting process, warp and weft silk scouring and dyeing process, silk preparation process and silk weaving process. In the weaving process, the first part is the pattern drafting process. It involves drawing a pattern grid and color setting in accordance with the number of weaving machine stitches, the warp and weft density of the fabric, the type of pattern design and the size of the pattern design. Next is the skein silk scouring and dyeing process, in which the raw silk is changed into boiled-off silk, namely that the raw silk is dyed. The third part is the preparation process of the weaving: the package of the warp and weft silk and the installation. The fourth part is the weaving on the machine. What' s more, Shu brocade is woven by jacquard loom.

二、杭罗织造技艺/Hang Leno Weaving Skills

杭罗织造技艺是浙江省杭州市地方传统手工技艺，是国家级非物质文化遗产之一。它有着悠久历史，早在宋朝地方志中就已有记载，至清朝已成为杭州丝绸的著名品种。杭罗织造技艺复杂、严格，随着织机的改革创新，但其生产流程中仍保持着大量精细缜密的手工技艺，对生产者的技术要求极高。准备织造的生丝必须经过严格的检验和筛选，在上机织造前，需要经过浸泡、晾干、翻丝、纤经、摇纡等系列工艺，然后才能上机织造。织成的粗坯还要经过精练、染色等工序，才能成为精致的杭罗。2009年，杭罗织造技艺作为中国传统桑蚕丝织技艺的重要子项目之一，被列入联合国教科文组织《人类非物质文化遗产代表作名录》。

Hang leno weaving skill, a traditional local handicraft technique in Hangzhou, Zhejiang Province, is one of the national intangible cultural heritages. It has such a long history that it was recorded in local chronicles of the Song Dynasty. Until the Qing Dynasty, it had grown as a famous variety of Hangzhou silk. Its weaving technique is so complex. There still exists a large amount of fine and meticulous craftsmanship coupled with changes in the loom. Therefore, the requirements for the producer’ s skills are extremely high, with its weaving process entailing painstaking efforts. Firstly, the raw silk filaments that are ready for weaving must undergo strict quality inspection and screening. Before woven on the machine, they need to be subject to a series of processes such as impregnation, drying, silk dyeing, thin filament, weft–shaking, etc. Then the woven rough flush must be refined and dyed to become exquisite Hang leno. In 2009, Hang leno weaving skill was inscribed by UNESCO on the *Representative List of the Intangible Cultural Heritage of Humanity*, one of the important sub–projects of traditional Chinese mulberry silk weaving skills.

杭罗绸面有等距的直条形或横条形纱孔，孔眼清晰，织法透气，质地光柔滑爽。这种面料穿着舒适凉快，耐穿耐洗，十分适合闷热、多蚊虫的夏天，广泛用作帐幔、夏季衬衫和休闲服面料等。在织法和工艺上，杭罗与其他丝绸也有明显的不同之处，在选材上，杭罗由纯桑蚕丝织造而成，比较好的桑蚕丝有杭嘉湖地区产的蚕丝，牢度高、粗细均匀、有光泽。在加工技术上，杭罗以平纹和纱罗组织联合构成，采用特有的“水织”技术，先浸泡、后织造，从原料蚕丝加工到摇纡织造蚕丝都浸泡在特殊的水溶液里，如此织出的杭罗柔软，优于现代技艺织法。

Hang leno feels soft, smooth and cool due to its uniform and clear voids and breathable texture. This kind of fabric is comfortable and cool to wear, durable and washable. Suitable for sultry and mosquito–prone summer, it has been widely used as a curtain, summer shirt and casual clothes fabric. In the selection of materials, there is also a marked difference in the material selection between Hang leno and other silks, namely that Hang leno is entirely made of mulberry silk. There is Hangjiahu plain silk, for example, with its high fastness, even thickness and lustre.

Hang leno is composed of plain weave and gauze and leno. When processing, the water weaving technique is adopted, which means impregnating first and then weaving. The silk is soaked in a special water solution from the processing of raw silk to the weft-shaking weaving. In this way, Hang leno is soft, superior to modern technical weaving.

三、潞绸织造技艺 /Lu Silk Weaving Skills

潞绸，即古潞州织造之绸，其起源无法考证，但有一点是可以肯定的：早在盛唐时期，山西的丝与绸就已在“丝绸之路”上进行贸易。它是山西丝绸业鼎盛时期的代表，产于山西省长治市，因潞州而得名，历史上曾与杭罗、蜀锦齐名，被认为是中国历史上三大名绸之一，是极富地方特色的特色传统丝织品。在明清两朝，它是皇室贡品，也是支撑晋商发展的主要商品。潞绸因其精致和美丽，在明朝成为受欢迎的出口商品。传统潞绸手感厚实、结实耐用，出土实物的组织结构均为三枚斜纹地和纬六枚斜纹提花，辅以刺绣、手绘等。2014年，潞绸手工织造技艺被列入国家级非物质文化遗产名录。

Lu silk, the woven silk in ancient Luzhou in Shanxi Province, is impossible to be tested as to its origin, but one thing is certain: Shanxi silk had been already in trade on the Silk Road as early as the grand Tang Dynasty. As the representative of the Shanxi silk industry, it was produced in Changzhi, Shanxi Province. It is regarded as one of the three famous silks in China in history (the other two: Hang leno and Shu Brocade). What' s more, it serves as a royal tribute in the Ming and Qing Dynasties and major commodity of Shanxi merchants. Lu silk, with its exquisiteness and beauty, became a sought-after export in the Ming Dynasty. The traditional Lu silk feels thick, strong and durable. Its texture of unearthed objects is 3 twill weave and 6 twill jacquard, coupled with embroidery and hand painting. In 2014, Lu silk hand-woven skill was included in the national intangible cultural heritage list.

自隋朝以来，潞绸成为山西州府进贡朝廷的主要物品，说明蚕桑在长治等地兴盛已久，并在当时形成了极其深远的影响。明朝潞绸曾发展到鼎盛时期，山西的潞州成为北方最大的织造中心。潞绸长期以来一直是朝廷的贡品。明朝中期以后，潞绸成为全国畅销的产品。明万历年间，创作的《醒世恒言》曾多处提及潞绸。在其他典籍中也有关于潞绸的记载和描写，这表明潞绸生产和销售的繁荣情况。

Since the Sui Dynasty, Lu silk has become a major item in Shanxi prefecture' s tribute to the court, suggesting that the sericulture flourished in Changzhi and other places and had a far-reaching influence at that time. In the Ming Dynasty, Luzhou became the largest weaving center in the north in its heyday. Lu silk had long served as a royal tribute to the imperial court. After the middle of the Ming Dynasty, it became a best-selling product nationwide. It is mentioned in the famous book *Aphorisms to Awaken the Public*, which was created during the Wanli period of the

Ming Dynasty. Other books also carry passages about Lu silk, an indication of the prosperity of Lu silk production and sales.

四、双林绫绢织造技艺 /Shuanglin Ghatpot Silk Weaving Skills

双林绫绢产自浙江省湖州市双林镇，其织造技艺是当地传统手工技艺，由于地处太湖文化、古运河文化、吴越文化的交融之中，因此有着深厚的文化底蕴。中国的传统丝织品中，绫居首位，双林绫绢，从唐朝起便被列为贡品并远销日本等地。杭嘉湖水网地带，历来盛产蚕桑，缫丝业相当发达。双林绫绢被誉为东方丝织工艺的奇迹，其生产历史悠久，源远流长。绫绢业的发展，使双林很早就成为江南水乡著名的商埠巨镇。2008年双林绫绢织造技艺被列入国家级非物质文化遗产名录。

A traditional local craftsmanship, Shuanglin ghatpot silk is from Shuanglin Town, southeast of Huzhou, Zhejiang Province. It boasts a deep-rooted cultural influence, situated in the mixture of Taihu Lake culture, ancient canal culture and Wu Yue culture. Silk twill ranking first among the traditional Chinese silk textiles, it was listed as a tribute since the Tang Dynasty and exported to Japan and other places. Located in the Hangjiahu water network area, it has always been rich in sericulture and the reeling industry is quite developed. Known as the wonders of oriental silk weaving craftsmanship, it has a long and profound history in production. It is the development of the ghatpot industry that has made Shuanglin a famous commercial giant town in canal towns in Southern China. In 2008, Shuanglin ghatpot silk weaving skills was included in the national intangible cultural heritage list.

绫绢是绫与绢的合称，“花者为绫，素者为绢”，用纯桑蚕丝织制而成。双林绫绢轻似晨雾、薄如蝉翼、质地柔软、色泽光亮，素有“凤羽”之美称，被誉为“丝织工艺之花”，生产历史非常悠久。双林绫绢传统手工技艺精湛，种类繁多，分为耿绢、矾绢、花绫、素绢等诸多品种，生产工艺流程严密，主要有：浸泡、翻丝、纤经、放纡、织造、炼染、研光、整理等20余道工序。

Ghatpot and lustre are made of pure silk. The difference between them lies in that the patterned one is ghatpot and the plain is silk twill. Known as the “phoenix feather”, and “the crown of the silk weaving process”, Shuanglin ghatpot silk is as light and thin as cicada wings, soft in texture and bright in color. With a long production history, the traditional Shuanglin ghatpot silk handicraft technique is exquisite and varied. This type of silk includes lustre, vitriol-dyed ghatpot silk, jacquard twill, plain silk and others. The production process flow is strict, mainly including more than 20 processes such as soaking, turning over silk, fiber warp, winding, weaving, refining and dyeing, calendaring, and finishing.

第四章 丝绸印染技艺 Silk Dyeing Techniques

第一节 蚕丝纤维的表征/Characterization of Silk Fiber

蚕丝是由蛋白质构成的，其中丝素（或称丝朊）主要成分是蛋白质，丝胶的主要成分是球状蛋白。当蚕在吐丝时，可以同时吐两根丝，由丝胶黏合为一根茧丝。丝胶在营茧时起黏合作用，使许多排列整齐的茧丝黏合起来而形成茧层。丝胶约占茧层重量的1/4，在茧外层较多，内层较少。茧层内还含有脂肪、蜡质、碳水化合物、色素及无机物等，重量约占茧层的3%，其中脂肪及胶质占0.4%~0.8%，碳水化合物占1.2%~1.6%，色素占0.2%，无机物占0.7%，主要集中于丝胶部分。蜡质在茧外层及茧内层较多，中层较少。以一根丝而言，茧丝表面集聚了部分蜡质，最接近丝素的丝胶中含蜡质较多，因此造成脱胶和精练困难。

Silk is made up of proteins. In silk, the main component of fibroin (or silk fibroin) is protein fiber, and the main component of sericin is globular protein. When silkworms produce the cocoon silk, they can produce two silks at the same time. At this point, the two parts are bonded by sericin into one part. Sericin can play a bonding role when the silkworm makes a silkworm cocoon, binding neatly arranged cocoon silk to form a cocoon layer. Accounting for about 1/4 of the total weight of the cocoon layer, it is distributed more in the outer layer and less in the inner layer. There are also fats, wax, carbohydrates, pigments, inorganic substances and so on within the cocoon layer. Their weight accounts for about 3% of the layer, of which fat and gum account for about 0.4% to 0.8%, carbohydrates 1.2% to 1.6%, pigments 0.2%, and inorganic substances 0.7%, mainly concentrated in the sericin section. The wax content is more in the outer layer and the inner layer of the cocoon, while that in the middle layer is less. Some wax on the surface of cocoon silk and more on sericin closest to fibroin causes trouble in degumming and refining.

蚕丝中的主要化学元素包括钙、镁、钠、钾，还有铝、锰、硫、磷等24种元素，这可能从饲养的桑叶中摄入，主要存在于丝胶中。

The main chemical elements in cocoon silk include calcium, magnesium, sodium and potassium, and 24 other elements, such as aluminum, manganese, sulfur and phosphorus. They may be ingested by mulberry leaves, which mainly exist in sericin.

丝胶和丝素的结构及其性质，对丝织物的精练、印染加工有重要的影响。

The structures and properties of sericin and fibroin have an important influence on the refining, printing and dyeing of silk fabrics.

一、丝胶的结构及性能 /Structures and Properties of Sericin

早在19世纪就开始进行丝胶及丝素化学结构的研究。1962年，日本桐村二郎确定了丝胶由19种 α－氨基酸组成，与丝素的 α－氨基酸组成基本相同，并定量地确定了19种 α－氨基酸的含量（表4-1）。

Research on the chemical structure of sericin and fibroin began as early as the 19th century. In 1962, Erro Tomura of Japan determined that sericin consists of 18 α–amino acid species. It is basically the same as the α–amino acid composition of fibroin and quantitatively determined the contents of 18 α–amino acids (Table 4-1).

表4-1　丝胶和丝素的α-氨基酸组成
Composition of α-amino Acid for Sericin and Fibroin

α–氨基酸名称/Alpha-amino acid name	在丝胶中的含量/Content in sericin (%)	在丝素中的含量/Content in fibroin (%)	侧链亲疏水性/Side chains with hydrophobicity or hydrophilicity
甘氨酸/Glycine	4.1～8.8	42.6～48.3	疏水/Hydrophobicity
丙氨酸/Alanine	3.51～11.9	32.6～35.7	
缬氨酸/Valine	1.3～3.14	3.03～3.63	
亮氨酸/Leucine	0.9～1.7	0.68～0.81	
异亮氨酸/L-isoleucine	0.6～0.77	0.87～0.9	
苯丙氨酸/Phenylalanine	0.5～2.66	0.48～2.6	
蛋氨酸/Methionine	0.1	0.03～0.18	
胱氨酸/Cystine	0.2～1	0.03～0.88	
酪氨酸/Tyrosine	3.77～5.53	11.29～11.8	稍亲水/Slight hydrophilicity
色氨酸/Tryptophan	0.5～1	0.36～0.8	
脯氨酸/Proline	0.5～3	0.4～2.5	
羟脯氨酸/Hydroxyproline	—	1.5	

续表

α-氨基酸名称/Alpha-amino acid name	在丝胶中的含量/Content in sericin (%)	在丝素中的含量/Content in fibroin (%)	侧链亲疏水性/Side chains with hydrophobicity or hydrophilicity
天门冬氨酸/Aspartic acid	10.43～17.03	1.31～2.9	亲水/Hydrophilicity
谷氨酸/Glutamic acid	1.91～10.1	1.44～3	
苏氨酸/L-threonine	7.48～8.9	1.15～1.49	
丝氨酸/Serine	13.5～33.9	13.3～15.98	
组氨酸/Histidine	1.1～2.75	0.3～0.8	
赖氨酸/Lysine	0.89～5.8	0.45～0.9	
精氨酸/Arginine	4.15～6.07	0.9～1.54	

从表4-1可看出，在丝胶中，侧链带亲水性基团的氨基酸（如丝氨酸、天门冬氨酸）含量最高，而在丝素中，侧链带非极性疏水性基团的氨基酸（如甘氨酸、丙氨酸）含量最高。因此，丝胶在水中的溶解性比丝素好，而丝素的结晶度、微晶取向度、强度等比丝胶要大得多。

As can be seen from Table 4-1, the content of amino acids with hydrophilic groups in side chains (such as serine and aspartic acid) is the highest in sericin. In fibroin, the content of amino acids (such as glycine and alanine) with non-polar hydrophobic groups in the side chain is the highest. Therefore, the solubility of sericin in water is better than that of fibroin. However, the crystallinity, microcrystal orientation and strength of fibroin are much larger than those of sericin.

1. 丝胶的结构/The Structures of Sericin

丝胶具有层状结构，可分为丝胶Ⅰ、丝胶Ⅱ、丝胶Ⅲ和丝胶Ⅳ。

Sericin has a layered structure and can be divided into sericin Ⅰ, sericin Ⅱ, sericin Ⅲ and sericin Ⅳ.

（1）丝胶Ⅰ。相对分子质量为31万，处于蚕丝的最外层，具有黏附性，质地硬而脆。染色以碱性染料为佳。共存的不饱和脂质的氧化反应会使溶解度降低，并加快泛黄速度。精练后有不均匀的残留，这是染色不匀和绞练丝黏着的原因。在与丝胶Ⅱ的交界处，存在一层与丝胶Ⅱ的混合层。

Sericin I. Its relative molecular mass is 310,000. It lies in the outermost layer of cocoon silk, with adhesion, hard and brittle texture. The alkaline dye is preferably used for dyeing. The oxidation reaction of the co-existing unsaturated lipid reduces solubility and speeds up yellowing. There will be uneven residue after refining, which is also the reason for uneven staining and adhesion of the twisted wire. At the junction with sericin Ⅱ, there exists a mixed layer with sericin Ⅱ.

（2）丝胶Ⅱ。相对分子质量为22万，该层丝胶表面平滑，具有良好的光泽。质地稍硬纤维刚性好，但形成织物后则无硬感，染色性良好。抗皱性、防缩性好，因反射或吸收紫外线而使产生的泛黄变褐速度缓慢。微茸埋入在此层丝胶内。

Sericin Ⅱ. Its relative molecular mass is 220,000. The surface of the sericin layer is smooth and has a good luster. Its texture is slightly harder, and it can give the fiber rigidity. It has no hardness when formed into a fabric, but it has good dyeability. It has good anti-crease and shrinkage resistance. Due to its reflection or absorption of ultraviolet light, the speed of yellowing and coarse cloth is slow. Lousiness is embedded in this layer of sericin.

（3）丝胶Ⅲ。相对分于质量为11万。分子间有S—S交联，但数量很少。因而不具备作为纤维材料的强度，没有丝胶Ⅱ那样的平滑表面，可用酸性染料染色。

Sericin Ⅲ. Its relative molecular mass is 110,000. There is S—S cross-linking between the molecules, but the number is small. Therefore, it does not have the strength of fibrous materials and the smooth surface of sericin Ⅱ. Acid dyes can stain well.

（4）丝胶Ⅳ。相对分子质量为17万。其表面平滑性与丝胶Ⅲ相似，含有较多的蜡质，很难溶解，因而耐水洗，以往的精练方法洗涤都有可能残留这种丝胶，可用碱性染料染色。

Sericin Ⅳ. Its relative molecular mass is 170,000. Its surface smoothness was similar to that of Sericin Ⅲ. Containing a large amount of wax, it is most difficult to be dissolved and therefore resistant to washing. Washing with previous refining methods may end up with residual sericin. Basic dyes can stain well.

2. 丝胶的性能/The Properties of Sericin

这四层丝胶的性能比较：

Properties of these four layers of sericin for comparison:

（1）易伸性。丝胶Ⅰ＜丝胶Ⅱ＜丝胶Ⅲ＜丝胶Ⅳ≤丝素。

Flexibility. Sericin Ⅰ < Sericin Ⅱ < Sericin Ⅲ < Sericin Ⅳ ≤ fibroin.

（2）染色性。丝胶Ⅰ是弱碱性蛋白质，与酸性蛋白质的丝胶Ⅱ和丝胶Ⅲ的染色性差异大。丝胶Ⅰ不均匀的残留是造成染色不均匀的原因。

Dyeability. Sericin I is a weakly basic protein that differs greatly from the Sericin Ⅱ and Sericon Ⅲ of acid proteins. The uneven residue of Sericin Ⅰ is the reason for uneven dyeing.

（3）脂质含量。有氧化反应结果的丝胶和化学反应的脂质，很难定量表示。

Content of lipid. It is difficult to quantitatively express the oxidized sericin and chemically reacted lipids.

（4）耐水洗性。丝胶Ⅱ层厚，脂质丰富，适当残留丝胶Ⅱ有利于洗涤。

Washability. Sericin Ⅱ is thick and rich in lipids, so proper residue of Sericin Ⅱ is beneficial to washability.

以往把丝胶作为废料处理，其溶液只能作为匀染剂使用。现在人们普遍认识到，从蚕茧加工成丝绸要经过多道工序，而在这些工序中，丝胶起着重要的作用。如果没有丝胶，绉绸就不会产生皱缩效应。加强捻的纬丝在热定形时靠丝胶固定；织物经精练后，原来被固定在茧丝内部的应力得以缓和，然后织物恢复到原来的状态，最后产生皱缩效应。纱罗织物和薄绸等夏季衣料，也可根据丝胶去除情况，获得类似于薄型平纹毛织物的衣料；含有适量丝胶，丝织物的挺阔性也会增加。此外，可把含丝胶40%～50%的蚕茧衣制成被褥，将有适度吸放湿性的丝胶用于涤纶等织物的丝胶改性加工中。

With sericin treated as waste in the past, its solution could only be used as a leveling agent. It is, however, generally recognized that sericin plays an important role in many processes of making silk from silkworm cocoons. Without it, crepe would not produce the wrinkling effect. In a heat setting, we can strengthen the twist of the weft wire fixed by sericin. After the woven fabric has been refined, the stress originally fixed inside the cocoon can be alleviated. The fabric can then be restored to the original state, and finally the wrinkling effect is generated. For summer fabrics, such as leno and chiffon fabric, we can also obtain fabrics similar to thin plain wool fabrics according to the removal of sericin. An appropriate amount of sericin contained, the crispness also increases. In addition, we can make cocoon outer floss with sericin about 40% to 50% into bedding, and use sericin with moderate moisture absorption and release to modify polyester fabrics.

3. 丝胶的其他化学性质 /Other Chemical Properties of Sericin

（1）黏着性。从蚕茧中直接分离出来的丝胶黏着性较差，但经适当的化学处理，可分离出类似动物胶的那部分丝胶，并能配制任意浓度的溶液，它具有吸湿性，会产生具有相当黏着性的物质。

Adhesion. Sericin that is isolated directly from silkworm cocoon has poor adhesion. But with proper chemical treatment, we can isolate the parts of sericin which are like animal glue and make solutions of any concentration. It is hygroscopic and would produce substances with quite adhesive properties.

（2）吸湿性。丝胶吸湿性能较好。在相对湿度为65%时，其回潮率为10.65%。相对湿度低于55%时，丝胶不发生变性。

Moisture absorption property. Sericin has a great hygroscopic effect. When the relative humidity is 65%, the moisture regain is 10.65%. Below 55% of relative humidity, sericin does not degrade.

（3）热分解性。丝胶较丝素易在低温下分解。丝胶含有较多的丝氨酸、天门冬氨酸、谷氨酸等，因而比丝素更容易在低温下分解。

Thermal decomposition. Compared with fibroin, sericin is easier to decompose at a low temperature, since it has more serine, aspartate, glutamic acid, and so on.

二、丝素的结构及性能 /Structures and Properties of Fibroin

丝素与丝胶都是由 α－氨基酸组成，但丝素的结构比丝胶更复杂。它是由不同的 α－氨基酸以肽键连结而形成的纤维大分子。其分子结构包含结晶区和非结晶区。其中，结晶区分子结合紧密，占丝素总量的40%～60%，非结晶区结构疏松，支链上的活性基团较多，这些基团易与染化料反应。可以利用丝素的这种特性，进行染色和后整理等染整加工。

Both fibroin and sericin are composed of α–amino acids, but their structures are more complex than those of sericin. It is a fibrous macromolecule formed by different α–amino acids through peptide bonds. Its molecular structure contains crystalline and non–crystalline regions. Among them, the molecules in the crystalline region are closely bound, accounting for about 40% to 60% of the total silk wrap. However, the structure of the non–crystalline zone is loose. There are more active groups on the branched chain, and therefore these groups are easy to react with dyeing. We can make use of this characteristic of silk fibroin for dyeing and finishing.

在标准环境中，丝素的回潮率为9.9%。丝素经长时间的煮沸，会有部分溶解。其丝素的等电点pH为2～3，在高浓度的中性盐溶液中，丝素蛋白质能溶解成溶液，且具有相当的可塑性。丝素对强碱和强酸的作用比较敏感，有些工厂在特定的温度下将强碱和强酸用作丝绸砂洗加工的砂洗剂，正是利用此特性。含氯氧化剂如次氯酸钠、亚氯酸钠、漂白粉等对丝素破坏作用极大，因此一般用双氧水作为蚕丝漂白水，或用如保险粉、亚硫酸钠等还原剂使蚕丝色素还原而脱色。

In a standard environment, the moisture regain of fibroin is 9.9%. Fibroin dissolves when boiled for a long time. With its isoelectric point pH at 2 to 3, fibroin dissolves into an aqueous solution in a high concentration of neutral salt solution. Besides, it has considerable plasticity. Fibroin is sensitive to strong bases and acids. Some factories use strong bases and acids at a specific temperature as a sand–washing agent for silk sand–washing processing. For chlorine–containing oxidants such as sodium hypochlorite, sodium chlorite, and calcium hypochlorite, the damage to fibroin is extremely great. Therefore, it is generally recommended to use hydrogen peroxide as a bleaching agent for silk or use reductants such as sodium hydrosulfite and sodium sulfite to reduce silk pigment and thus decolorize.

第二节　前处理 /Pre–processing

蚕丝除了丝胶和丝素外，还有蜡质、色素和灰分等杂质，此外在泡丝、制织和捻丝过程中，又会沾上第二次杂质。这些杂质和不纯物不仅会成为染色的障碍，而且会影响蚕丝和成品的质量，因此，必须在染色前将其去除。去除色素以外的杂质的工艺过程称为精练；

去除色素的工艺过程称为漂白。

In addition to sericin and fibroin, there are impurities such as wax, color and ash in cocoon filaments. In the process of soaking silk, weaving and twisting, the filaments will be stained again with impurities. These impurities and impure materials not only become an obstacle to dyeing, but also influence the quality of silk and finished products. Therefore, these must be removed before dyeing. The process of removing impurities other than pigments is called refining, and the process of removing pigment is called bleaching.

一、精练 /Refining

蚕丝精练的目的就是去除蚕茧和蚕丝织物上的丝胶、杂质、色素等物质，使蚕丝纤维呈现出天然的优良特性，同时使织物纹路清晰，渗透性能良好，便于进一步印染加工。

The purpose of silk refining is to remove the sericin, impurities, pigments, and other substances on cocoon and silk fabrics. Finally, it makes the silk fiber more natural, the texture of the fabric clearer and the penetrability better, facilitating printing and dyeing.

蚕丝的精练说到底就是丝胶的溶解和去除。丝胶与丝素是蛋白质，但它们的氨基酸组成比例不同，分子结构有很大的差别。因此可利用它们的差异，采用各种方法把丝胶从丝素上剥落下来而不损伤丝素。蚕丝的精练有以下几种常用的方法。

In the final analysis, silk refining is the dissolution and removal of sericin, which is the same protein as fibroin. However, their amino acid composition ratio and their molecular structure vary greatly. Taking advantage of their differences, we can use various methods to finally peel the sericin off the fibroin without damaging them. There are several commonly used methods to refine silk.

1. 热水精练 /Hot Water Refining

在缫丝时，将蚕丝置于50 ~ 60℃的热水浴中，有1% ~ 3%的丝胶溶于水中。将此生丝进而放在热水中，则丝胶继续逐渐溶出。如在115℃的热水中用高压釜处理180分钟，约96%的丝胶溶出。为了完全去除丝胶，必须用120℃的热水进行4次2小时萃取。所以，如果用热水来脱胶，对能源消耗和生产效率来讲都是很不经济的，而且丝素经长时间高温处理也会部分水解而失去光泽。再者，丝纤维和织物还有许多油脂、蜡质，清水不易将其浸润渗透，因而，在实际生产上，用热水脱胶是不现实的。如果进行脱胶预处理，需将丝胶预先膨化，但不如添加表面活性剂的效果好。

When reeling silk is processed, cocoon filaments are placed in a hot water bath of 50℃ to 60℃, and 1% to 3% of the sericin will dissolvein water. Raw silk is put in hot water, and the sericin gradually dissolves. If treated with an autoclave in hot water at 115℃ for 180 min, about 96% of its sericin dissolves. To completely remove the sericin, we must use hot water at 120℃

to make extraction 4 times in 2 hours. Therefore, it is uneconomical for energy consumption and production efficiency to degum with hot water, and silk fibroin will be partially hydrolyzed and lose its luster after long-term high-temperature treatment. At the same time, silk fibers and fabrics still have a lot of grease and wax, which are not easily infiltrated by clean water. Therefore, degumming with hot water is unrealistic in actual production. If it is pretreated by degumming, it is possible that sericin is expanded in advance. However, its effect is not as good as that of adding surfactant.

2. 酸精练 /Acid Refining

从理论上讲，蚕丝在pH为1.75 ~ 2.5的范围内可以脱胶。有机酸对丝胶蛋白的影响很小，只有在强无机酸作用下，丝胶蛋白才会发生水解。但强酸容易破坏丝素，而且酸去除油蜡等杂质的能力也很低。据文献报道，有一种工艺用于绢纺原料及生丝半脱胶的工艺，其步骤为：①先将生丝用1克/升硫酸溶液在95 ~ 97℃短时间浸渍，使脱胶率达到10% ~ 12%；②用纯碱溶液中和后；③进行皂练。但用酸脱胶的成品，手感发硬发糙，而且可能对丝素造成损伤，工艺复杂。此外用酸处理需要耐腐蚀的设备，因而，在实际生产中，很少使用酸来对丝织物进行精练。

Theoretically, silk can be degummed in the pH range of 1.75 to 2.5. Organic acids have little effect on sericin protein. Under the action of strong inorganic acid, sericin protein will be hydrolyzed. However, strong acid easily damages fibroin, and the efficiency of acid to remove impurities such as oil wax is also low. According to the literature, there is a process that can be used for semi-degumming of silk spinning of raw materials and raw silk, which involves: ① impregnating the raw silkwith 1 gram or 1 liter of sulfuric acid solution for a short timeat 95 to 97℃ to make degumming reaches 10% to 12%; ② neutralizing it with soda solution; and ③ soaping it. However, the finished product degummed with acid makes it feel hard and rough, and it may damage fibroin. Its process is also complicated. In addition, we need corrosion-resistant equipment if it is treated with acid. Therefore, in practical production, we rarely use acid to refine silk fabrics.

3. 碱精练 /Alkali Refining

各种碱剂可以作为蚕丝的脱胶剂，包括氢氧化钠、碳酸钠、碳酸钾、硅酸钠、磷酸三钠、焦磷酸钠、碳酸氢钠、硼砂、氨水等。碱脱胶的原理是，在碱性溶液中，丝胶分子中羧基上的氢离子（H^+）会与碱溶液中的氢氧根离子（OH^-）结合，此外，丝胶蛋白质分子上的羧基会与钠离子（Na^+）结合生成蛋白质钠盐，而加速其溶解。

Various alkali agents can be used as degumming agents for silk, which including sodium hydroxide, sodium carbonate, potassium carbonate, sodium silicate, trisodium phosphate, sodium bicarbonate, borax, aqueous ammonia, and so on. The principle of alkali degumming is that in an alkaline solution, hydrogen ions (H^+) on carboxyl groups in sericin molecules will combine with hydroxide ions (OH^-) in alkaline solution. In addition, carboxyl groups on protein molecules of sericin itself combine with sodium ions (Na^+) to form protein sodium salts, which accelerating its dissolution.

当OH^-的浓度很高时，丝胶分子间的结合力减弱。此时，丝胶不断地与碱结合，膨化、剥落，进一步水解。同时，溶液中的氢氧离子也会随之而减少，即碱被丝胶吸收，而使碱溶液的碱性降低。

When the concentration of OH^- is very high, the binding force between sericin molecules will weaken than before. At this time, sericin will be continuously combined with alkali agents, expanded and peeled off, and further hydrolyzed. At the same time, the hydroxideion in the solution will also decrease. That is to say, the alkali agent is absorbed by sericin, which reducing the alkalinity of the alkali solution.

碱精练是常用的脱胶方法，常用碱剂的pH见表4-2。它们的脱胶速率次序为：烧碱>碳酸钠>硅酸钠>磷酸三钠>硼砂>碳酸氢钠。其中，碳酸钠是较佳的碱剂。

Alkali refining is the most commonly used degumming method. The pH of common alkali agents are shown in Table 4-2. Their degumming rates are as follows: caustic soda > sodium carbonate > sodium silicate > trisodium phosphate > borax > sodium bicarbonate. Among them, sodium carbonate is a better alkali agent.

表4-2　常用碱剂的pH
pH of Common Alkali Agents

碱剂/Alkali agent	pH（浓度为1%）/ pH value (concentration: 1%)
氢氧化钠/Sodium hydroxide	13.3
硅酸钠/Sodium silicate	12.4
磷酸三钠/Trisodium phosphate	12.1
碳酸钠/Sodium carbonate	11.5
硼砂/Borax	9.0
碳酸氢钠/Sodium bicarbonate	8.3

经验表明，在碱精练时，精练浴的pH宜为8.8 ~ 10.5。pH低于8.8，脱胶速率太慢，可能会发生机械损伤；pH高于10.5，化学损伤的可能性迅速提高。

During alkali refining, our experience shows that the pH of the scouring bath should be 8.8 to 10.5. If the pH is lower than 8.8, the degumming rate is too slow, which may lead to mechanical damage. If the pH is higher than 10.5, the possibility of chemical damage will increase rapidly.

4. 皂精练/Soap Refining

肥皂是丝绸脱胶、精练的主要助剂。在水中呈离子状态，属阴离子型表面活性剂。

Soap is the main auxiliary agent for degumming and scouring silk. Ionic in water, it belongs to an anion surfactant.

肥皂具有湿润、乳化的作用，去污力较强。用肥皂进行精练，作用比较温和，其水溶液pH在9～10，正好适用于蚕丝纤维的脱胶。丝胶可与肥皂中的钠离子相互作用。当溶解进入练液，肥皂中的一部分脂肪酸会被丝织物吸附，另一部分又被其余的钠离子补充，因而具有相当大的缓冲作用。对于织物上的油蜡杂质，肥皂也可将其乳化，使其分散进入练液中。用肥皂精练脱胶的成品因吸附微量油脂，光泽明亮，手感柔软滑爽，富有弹性。然而，皂精练也有较多缺点：①对蚕丝吸附强，难以用水洗掉；②在蚕丝纤维中残留的肥皂会带来染斑、泛黄、脆化等；③肥皂还易受精练液中硬水的影响。此外，皂精练工艺过程中肥皂耗用量大，精练时间长，不太经济，因此通常将肥皂和纯碱等碱剂混合进行精练。

Soap has the function of wetting and emulsifying, so it has strong detergency. We use soap for refining, and its effect is mild. The pH of its aqueous solution is between 9 to 10, so it is just suitable for degumming silk fibers. Sericin can interact with sodium ions in soap. When the solution dissolves, a part of the fatty acid in the soap will be absorbed by the silk fabric, while the other part will be supplemented by the rest of the sodium ions. Thus, it has a considerable buffering effect. For the oil wax impurity on the fabric, soap can also emulsify it, so as to disperse it into the scouring liquor. The finished product which is refined and degummed with soap will absorb trace oil. Therefore, the silk is bright, soft, smooth, and elastic. However, soap refining also has many disadvantages: ①it has strong adsorption to silk, making it difficult to wash off with water; ②in silk fiber, the residual soap causes staining, yellowing, tender and others; and ③soap is also susceptible to hard water in scouring solution. In addition, in the process of soap refining, the soap consumption is large, and the refining time is long, it is therefore uneconomical. As a result, we usually mix soap with alkali agents such as soda ash and refine them.

皂碱法精练是以肥皂作主练剂，以纯碱、硅酸钠等作碱剂，保险粉作还原性漂白剂。该方法中实际上起主要作用的还是“碱”。精练液的pH维持在9.5～10.5的范围内，丝胶才能快速溶解或水解。肥皂可以使丝胶膨化、渗透，并凭借其乳化能力和去污能力，将织物上的丝胶和油蜡杂质乳化分散到练液中。硅酸钠除了提供碱性钠离子外，也能水解成硅酸。硅酸是不溶于水的白色凝胶，具有保护胶体的作用，可有效防止已剥落到练液中的丝胶杂质再度沾到织物上去；它还能吸附水中的杂质及铁离子、铁的氧化物，从而起到除锈作用，有助于提高成品的白度。保险粉是强还原剂，在碱性液中它能破坏丝纤维上的色素及织造过程中染料着色。

The soap-alkali method of refining uses soap as the main refining agent, soda ash and sodium silicate as alkaline agents, and sodium hydrosulfiteas a reducing bleaching agent. In fact, the main role in this method is “alkali” . If the pH of the refining solution is kept within the range of 9.5 to 10.5, sericin can dissolve or hydrolyze quickly. Soap can make sericin expand and permeate, and by virtue of its emulsifying power and decontamination ability, sericin and oil wax impurities on fabric are emulsified and dispersed into the scouring liquor. Besides providing basic sodium ions, sodium silicate can also be hydrolyzed into silicic acid. Silicic acid is a white gel that is insoluble in water and acts as a protective gel. It can effectively prevent sericin impurities peeled off from the scouring liquor from sticking to the fabric again. It can also adsorb impurities, iron ions and iron oxides in water so as to play a role in rust removal. It helps to improve the whiteness of finished products. Sodium hydrosulfite is a strong reducing agent. In alkaline scouring liquor, it can destroy the pigment on silk fibers and the dyed dyes during weaving.

5. 酶精练/Enzyme Scouring

酶精练是将蛋白质分解酶应用于蚕丝精练的方法，它又叫“生物精练”，或利用生物技术提纯蚕丝。

Enzyme scouring, a method of applying protease to silk refining, is also known as “biological refining” , or the use of biotechnology for silk refining.

蚕丝采用酶精练，不仅可以在较低的温度下处理，使丝素不受损，还可获得蓬松性好、不起毛等优良品质，同时也能去除死茧造成的污垢。

Refined by enzyme scouring, the silk can be treated at a lower temperature so that the fibroin cannot be damaged. Refined silk with enzyme scouring has the characteristics of good bulkiness and hairlessness, and it can also remove dirt caused by dead cocoons.

酶脱胶一般由以下三道工序组成：首先对纯碱和表面活性剂进行预处理，使丝胶膨润、软化，从而使酶容易发挥作用；然后进行酶处理，使生丝丝胶分解、去除；最后进行后处理，以洗去附着在蚕丝上的成分，从而得到脱胶均匀的蚕丝。

Enzymatic degumming generally consists of the following three processes. First, we pretreat soda ash and surfactant to make sericin swell and soften. The enzyme is thus easy to play its role. When enzyme treatment is carried out, sericin decomposes from raw silk. When post-treatment is conducted, the components attached to the silk are washed away, and thus the degummed silk is obtained.

研究表明，蛋白质分解酶在多肽基质的有限数量的位置上反应。酶的这种专一的作用方式，对丝素不起作用，仅对丝胶起作用，使经过处理的蚕丝纤维表面保持光泽优雅、很少起茸毛的优美状态。表4-3为皂碱法精练与酶精练练减率、外观和风格的比较。由表中数据可知，蚕丝的酶精练与皂碱法精练所得的结果很相近。

Studies have shown that proteases react at a limited number of positions in the polypeptide matrix. This specific mode of action of the enzyme has an effect on sericin instead of fibroin. In addition, it can make the surface of the treated silk fiber maintain a beautiful state with elegant luster and little fuzz. Table 4–3 shows the comparison of alkali scouring rate, appearance and style between the soap–alkali method and enzyme scouring. According to the data in the table, the results of enzyme scouring of silk are very similar to those of the soap–alkali method.

表4–3 皂碱法和酶精练法结果对比
Comparison of Scouring Results between Soap-alkali Method and Enzyme Scouring

脱胶方法/Degumming methods	皂碱精练/Soap–alkali method	酶精练/Enzyme scouring
练碱率/Alkali scouring rate (%)	26.37	29.62
沉淀吸附量/Precipitation adsorption capacity	多而大/ Much and large	没有/ Without
精练后光泽/Glossy after refinement	良（白）/Good (white)	优/Excellent
风格/ Style	弱/ Weak	柔（稍有紧密感）/ Soft (slightly tight)

蚕丝精练所用的酶主要有以下几种：

The main enzymes used in silk refining are as follows:

（1）胰蛋白酶。它是一种丝氨酸蛋白酶，从猪胰脏中提取。胰酶内同时含有淀粉酶和蛋白酶，在提取时以蛋白酶为主要成分的称为胰蛋白酶，以淀粉酶为主要成分的称为胰淀粉酶。胰蛋白酶的外观为微黄色的粉末，有腥味，最大活性范围在pH为7 ~ 9。碳酸氢铵（0.1mol/L）是较好的缓冲剂。酶 / 基质比率为1%或2%，宜在37℃时处理1 ~ 4小时。

Trypsin, a kind of serine protease, is extracted from the pig pancreas. Pancreatin contains both amylase and protease. In the extraction, protease is the main component called trypsin, and amylase is the main component called pancreatic amylase. The appearance of trypsin is a yellowish powder, which has a fishy smell. Its maximum activity range is between pH 7 and 9. Ammonium bicarbonate (0.1 mol/L) is a good buffer. The ratio of enzyme to substrate is 1% or 2%, which is suitable for treatment at 37℃ for 1 to 4 hours.

据国外有关报道，与仅含碳酸钠和蛋白酶的液体脱胶相比，使用含4%（质量分数）的碳酸氢钠、碳酸钠和硫酸钠的溶液，以2：2：1（质量比）的混合物和0.3%（质量分数）碱蛋白酶所组成的溶液，在50 ~ 60℃下处理10 ~ 20 min，最终能获得较好的脱胶效果。此外，用1%胰蛋白酶液，37℃处理10h，丝胶能完全水解。用碱预处理后，在含有2%胰蛋白酶 / 脂肪酶 / 淀粉酶以72：10：18（质量比）的比例混合的溶液中，pH为7 ~ 8。37 ~ 40℃处理1.5 ~ 2h，不仅能脱胶，还能去除脂肪和其他杂质。

According to foreign reports, compared with liquid degumming containing only sodium carbonate and protease, we use a solution containing 4% (by weight) sodium bicarbonate, sodium carbonate and sodium sulfate, a mixture of 2 : 2 : 1 by weight and 0.3% alkaline protease. And we treat it at 50℃ to 60℃ for 10 to 20 min, and finally we can get a better degumming effect. In addition, it was reported that 1% trypsin solution was used and treated at 37℃ for 10h to completely hydrolyze sericin. After we have pretreated it with alkali, then we can put it in the environment with a pH 7 to 8 and make it at 37℃ to 40℃ for 1.5 to 2h in the solution containing 2% (by weight) trypsin/lipase/amylase in the ratio of 72 : 10 : 18 by weight. Finally, we can degum it, and also remove its fat and other impurities.

（2）木瓜蛋白酶。木瓜蛋白酶在7～90℃、pH 5～7.5条件下活性最高。在精练过程中，开始温度保持在70℃以上，此时作用较强烈，随着温度下降，作用趋于缓和，这有利于保护丝素不受损伤。木瓜蛋白酶在硫氰酸盐、硫化氢、重亚硫酸钠、保险粉中活性较好。

Papain can exhibit maximum activity at 7℃ to 90℃ and pH 5 to 7.5. In the process of refining, the starting temperature should be kept above 70℃, at which time its effect is stronger. With the decrease in temperature, its effect tends to moderate. This is beneficial for protecting fibroin from damage. Papain has better activity in thiocyanate, hydrogen sulfide, sodium bisulfite, and sodium hydrosulfite.

木瓜蛋白酶对丝胶作用的机理初步认为，木瓜蛋白酶对多肽（缩多氨酸）的作用有广泛的专一性。木瓜酶可以有效地作用于精氨酸和赖氨酸残基的羧酸基形成的肽键。此外，木瓜蛋白酶在组氨酸的羧酸基上也迅速分裂，此分裂还发生在甘氨酸、谷氨酸、谷酰胺、亮氨酸和酪氨酸上。

We preliminarily think that the mechanism of papain's action on sericin is the effect of papain on polypeptide (peptide) which has a broad specificity. Papain can effectively act on peptide bonds formed by carboxylic acid groups of arginine and lysine residues. In addition, it can split rapidly on the carboxylic acid group of histidine. This division also occurs in glycine, glutamic acid, leucine, and tyrosine.

用木瓜蛋白酶精练蚕丝方法有：①一夜精练法。将丝织物生坯绸在木瓜蛋白酶溶液中浸渍一夜（15～16h），次日取出脱胶。②轧卷法。用非离子表面活性剂诺衣近HC进行前处理，然后授轧木瓜蛋白酶、保险粉、湿润剂等，温度75℃，一浸一轧。轧液率为100%，然后上卷，再用塑料袋包好，在75℃的恒温室内放置一定时间。最后，水洗，用诺衣近HC后处理，烘干。③快速脱胶法。采用平幅连续工艺，将丝绸在70℃温水中前处理3min后，放入70℃的木瓜蛋白酶及保险粉溶液中，浸渍3min后，放入70℃和湿度为85%～95%的精练室中精练。接下来，在含非离子表面活性剂的70℃温水中洗涤，然后水洗，再烘干。

Silk can be refined with papain in the following ways: ①Overnight refinement method. The method is to put the raw silk fabric in papain solution impregnated overnight (15 to 16 h), and take it out the next day for degumming. ②Rolling method. First, we can use nonionic surfactant which is called Nuoyijin HC for pretreatment. Then, we roll papain, sodium hydrosulfite, wetting agents, and so on, while the ambient temperature should be at 75℃ and the picked-up ratio is 100%. Next, we put the chemical modification volume in it. Then, we wrap it in a plastic bag, and then put it in a constant temperature room at 75℃ for a certain time. Finally, we can wash it with water, and treat it with surfactant and finally dry it. ③Rapid degumming method. We can use the open-width continuous process to operate it. At this point, the silk can be placed in warm water at 70℃ for 3 minutes before treatment. Then we put it into the solution of papain and sodium hydrosulfite at 70℃. After soaking it well for 3 minutes, we refined it in a scouring chamber at 70°C and humidity of 85% to 95%. Next, we can wash it in warm water containing nonionic surfactant at 70°C. Then, we can wash it with water and finally dry it.

（3）细菌酶。蚕丝也可以通过细菌酶来进行精炼。目前国内采用的细菌酶有2709碱性蛋白酶（属枯草杆菌类细菌蛋白酶）、209碱性蛋白酶（芽孢杆菌蛋白酶）、S114中性蛋白酶、ZS742中性蛋白酶和BF7658淀粉酶等。表4-4为S114中性蛋白酶和2709、209碱性蛋白酶对丝胶精练效果的比较。明显地，2709碱性蛋白酶对丝胶的分解能力比其他酶更强。

Bacterial enzymes. Silk can also be refined by bacterial enzymes. At present, the bacterial enzymes commonly used in China are 2709 alkaline proteases (belonging to Bacillus subtilis), 209 alkaline proteases (Bacillus protease), S114 neutral proteases, ZS742 neutral proteases and BF7658 amylases, etc. Table 4-4 shows the comparison of refining effects of S114 neutral proteases and 2709 and 209 alkaline proteases on sericin. Obviously, we can see that 2709 alkaline proteases have a stronger ability to decompose sericin than other enzymes.

表4-4 不同蛋白酶、不同时间精练情况
Refinement of Different Proteases and Different Times

酶种/Enzymes	浓度（活力单位）/ Concentration (activity unit, mL)	缩减率/Shrmkage（%）			
		30 min	60 min	120 min	180 min
S114中性蛋白酶/S114 ' neutral proteases	40	7.5	7.65	9.75	10.25
	60	9.0	9.45	11.15	11.80
209碱性蛋白酶/209 alkaline proteases	40	—	11.85	14.85	—
	60	—	13.60	16.25	—

续表

酶种/Enzymes	浓度（活力单位）/ Concentration (activity unit, mL)	缩减率/Shrmkage（%）			
		30 min	60 min	120 min	180 min
2709碱性蛋白酶/2709 alkaline proteases	40	12.75	16.0	19.45	19.45
	60	13.60	16.9	19.40	20.20

二、漂白 /Bleaching

蚕丝纤维的大部分色素存在于丝胶中。当丝胶去除后，织物白度得到改善。此外，丝素中还有一部分色素和织造时后施的着色染料，已在精练时被练液中的保险粉一起还原破坏，实际上精练和漂白（指还原漂白）是同时进行的。一般丝织物经第一次和第二次的精练还原漂白后，已经相当白。采用合成洗涤剂精练的白度更好一些，在白度计上测其白度可达85%。而且，丝织物也不必漂得太白，太白则失去丝织物的韵味。还原漂白的丝织物容易泛黄。为了降低丝织物的泛黄程度，或者对白度要求高的产品，特别是厚重不易练白的织物，可以采用氧化漂白。

Most of the pigments of silk fiber are present in sericin. When we remove sericin from silk, the whiteness of the fabric will be improved. In addition, there is a part of the pigment in the silk fibroin. When the fabric is woven, it is placed in the dyeing department later, which has been reduced together with the sodium hydrosulfite in the scouring liquor during refining and then destroyed. Refinement and bleaching (referring to the restoration of bleach) are carried out simultaneously. A silk fabric will be refined for the first and second time and it will already be quite white after two reduction bleaches. Therefore, the whiteness of scouring with synthetic detergent is better. Its whiteness can reach 85% measured by a whiteness meter. Moreover, the silk fabric need not be bleached too white. If the silk fabric is too white, it will lose the charm of the silk fabric. For reductively bleached silk fabrics, it turns yellow very easily. In order to reduce the yellowing degree of silk fabrics, or products with high white degree requirements, especially heavy and difficult to white fabric, we can use the method of oxidative bleaching.

所谓氧化漂白，也不是单纯用氧化剂漂白，而是采用还原漂—氧化漂—还原漂工艺，才能起到理想的效果。在漂白过程中，使用的氧化剂是过氧化氢。在酸性介质中，过氧化氢的性能较为稳定，其分解速度随pH及温度的提高而加速，铁离子等重金属离子也能促进过氧化氢的分解。因此，在漂液中往往加入稳定剂如硅酸钠、焦磷酸钠等，以防止重金属离子对双氧水的催化作用。一般常用硅酸钠，用量为1～2克/升，使漂液pH在9左右，另外加入平平加O等表面活性剂帮助渗透和扩散。过氧化氢（30%）的用量为1.5～2毫升/升，温度为80～90℃，处理1～2小时，所得成品白度比较稳定，能保持在83%～86%的水平，对泛黄程度也有所改善。

Oxidative bleaching does not refer to the use oxidants only to bleach fabrics. Instead, the method adopted is the process of reduction bleaching and oxidation bleaching and reduction bleaching process. Only in this way can a desired effect be achieved. In the bleaching process, the oxidant we use in most cases is hydrogen peroxide. And in an acidic medium, the performance of hydrogen peroxide is relatively stable. The decomposition rate will accelerate with the increase of pH and temperature. Heavy metal ions such as iron ions can also promote the decomposition of hydrogen peroxide. Therefore, stabilizers such as sodium silicate and sodium pyrophosphate are often added to the bleaching solution to prevent the catalysis of heavy metal ions on hydrogen peroxide. We usually use sodium silicate and its dosage is 1to 2 g/L so that the pH of the bleaching solution is about 9. In addition, we can also add surfactants such as Peregal O to help penetration and diffusion. The dosage of hydrogen peroxide (30%) is about 1.5 to 2 ml/L. The ambient temperature should be kept at 80 to 90℃, and the whiteness of our finished product is relatively stable after treating it for 1 to 2 hours. It can be kept at the level of 83% to 86% and the yellowing is also improved.

第三节　染色 /Dyeing

在漫长的历史中，曾用于染蚕丝的植物染料，有靛蓝、栀子、紫根草、青茅草、杨梅、苏木等3000多种；曾用于染蚕丝的动物色素，有贝紫、胭脂红等。还有用于染黄色的矿物色素。这些都为天然染料。随着化学染料的蓬勃发展，天然染料一度衰落。近年来，随着人们追求健康、渴望回归自然的希望日益迫切，天然染料特别是植物染料越来越受到人们的关注。

In a long history, people used to dye silk plants, including more than 3,000, such as indigo, cape jasmine, grassroots of a plant called Zicao, a kind of grass called Qingmao, red bayberry and logwood. People have used animal pigments for silk, such as a purple dye made from the gland of shellfish and carmine. And of course, there are mineral pigments used for yellow dyeing. These are all-natural dyes. With the development of chemical dyes, natural dyes were once ignored. In recent years, with the increasing urgency of people's health and desire to return to nature, natural dyes, especially plant dyes, have attracted more and more attention.

人工合成的化学染料，包括直接染料、酸性染料、含金属染料、金属媒染染料（铬媒染料）、盐基性染料、还原染料、活性染料等。蚕丝纤维几乎对所有的染料都有很高的亲和性。

Synthetic chemical dyes include direct dyes, acid dyes, metal-containing dyes, metal-medium dyes (chromium–medium dyes), salt–based dyes, reductive dyes, reactive dyes, and so on. Silk fibers have a high affinity for almost all dyes.

一、酸性染料染色 /Dyeing with Acid Dyes

酸性染料可用于蚕丝织物染色，染色方便，得色深浓、色谱齐全，价格低廉。酸性染料因染色性能和使用方法不同，可分为强酸性染料和弱酸性染料。强酸性染料分子量小，溶解度大，在溶液中会发生电离，与丝纤维亲和力较低，需pH控制在2～4时，才能使丝纤维上色。匀染性好，但色牢度差，多用于羊毛染色，丝织物不常采用。弱酸性染料分子量大，结构比较复杂，溶解后呈胶体溶液，和丝纤维的亲和力高。根据每种染料的性能不同，可在弱酸浴（pH为4～6）或中性浴中染色，染色牢度也较好。对丝织物染色，主要采用酸性染料。

Acid dyes can be used to dye silk fabrics, which have the advantages of convenient dyeing, deep color, complete chromatography, and low price. Due to different dyeing properties and application methods, acid dyes can be divided into strong acid dyes and weak acid dyes. Strong acid dyes have low molecular weight and high solubility, so they will undergo ionization reactions in solution and have a low affinity for silk fibers. Therefore, we need to control its pH at 2 to 4 to make silk fiber color. At this time, its leveling property is good, but its color fastness is poor, so this situation is mostly used in wool dyeing, and silk fabrics are not often used. Weak acid dyes have large molecular weights and complex structures. After dissolving, it is a colloidal solution, and its affinity for silk fibers is high. According to the different properties of each dye, we can dye the fabric in a weak acid bath (pH 4 to 6) or a neutral bath. Its dyeing fastness is also good, so we mainly use this kind of dye when dyeing silk fabrics.

1. 酸性染料的结构 /The Structures of Acid Dyes

酸性染料主要为芳香族有色化合物的磺酸钠盐。一般来说，它仅能在酸性介质或中性浴中对蚕丝等蛋白质纤维、聚酰胺纤维和酪素纤维进行染色，而对纤维素纤维无直接性。酸性染料的化学成分多为单偶氮或多偶氮化合物，约占此类染料的一半，其余为三芳甲烷结构及蒽醌结构。

Acid dyes are mainly sulfonic acid and sodium salt of aromatic and colored compounds. Generally, it can only dye protein fiber such as silk, polyamide fiber and casein fiber in a acidic medium or neutral bath, while it has no direct dyeing property for cellulose fiber. The chemical composition of acid dyes is mostly mono–azo or azo compounds. It accounts for about half of this kind of dye, while others are triaryl methane structure and anthraquinone structure.

下面列举了一些适用于蚕丝染色的弱酸性染料结构。

Some of the structures of weak acid dyes suitable for silk dyeing are listed below.

（1）单偶氮结构。卡普仑桃红 BS：

Structure of mono azo. Capron pink BS:

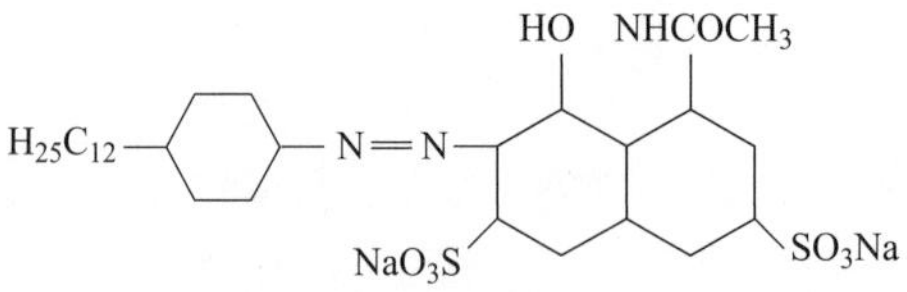

（2）双偶氮结构。弱酸性藏青 5R：

Structure of double azo. Weak acid navy blue 5R:

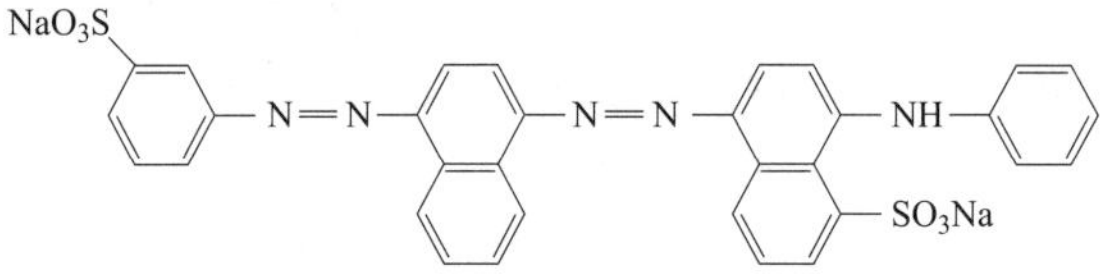

（3）三芳甲烷结构。弱酸性艳蓝 6B：

Structure of triaryl methane. Weak acid brilliant blue 6B:

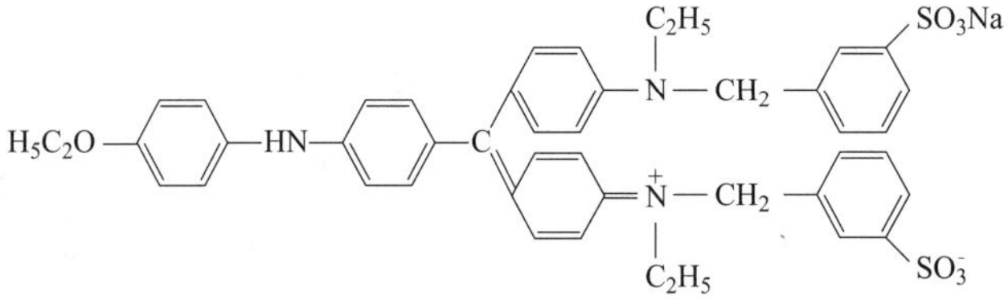

（4）蒽醌结构。卡普仑蓝 BS：

Structure of anthraquinone. Capron blue BS:

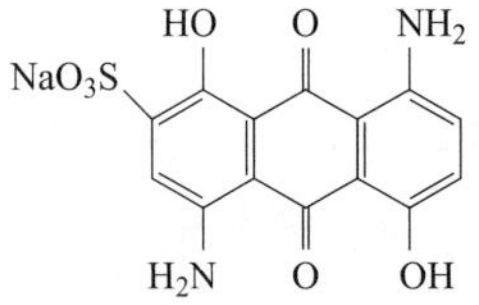

2. 酸性染料的染色机理 /Dyeing Mechanism of Acid Dyes on Silk

蚕丝纤维非结晶区蛋白质分子末端各类氨基酸上存在氨基、羧基、酚基等。丝纤维具有两性性质，在等电点以下的酸性溶液中，它呈阳荷性，而酸性染料在酸性中呈阴荷性，因而可利用酸性染料通过离子键上染到丝纤维上（“S”代表丝纤维分子，“D”代表染料分子）。

There are amino, carboxyl, phenolic groups, and so on in various amino acids in the terminal amino acids of protein molecules in the non-crystalline region of silk fiber. Silk fibers have amphoteric properties, which are positive in acid solutions below isoelectric points, but negative in acid dyes. Therefore, we can dye it onto silk fiber through its ionic bond by using acid dyes in this way(“S” stands for silk fiber molecule, “D” for dye molecule).

$$S\langle^{NH_2}_{COOH} \xrightarrow{H^+} S\langle^{NH_3^+}_{COOH} + D{-}SO_3^- \longrightarrow S\langle^{NH_3^+ \cdot D-SO_3^-}_{COOH}$$

弱酸性染料染色可在弱酸性浴中进行，也可在中性浴中进行。在弱酸性浴中染色时，一般pH应控制在4～6，这时因在丝纤维的等电点附近，所以有一部分染料还是借离子键与丝纤维结合，但也有一部分染料，因染液中没有足够的氢离子使丝纤维带有正电荷，而是借氢键与范德瓦耳斯力上染的。当在中性染液中染色时，染浴pH控制在6～7，这时染浴pH大于丝纤维等电点，染料和丝纤维的结合主要是靠分子间的碰撞，借氢键和范德瓦耳斯力上染。

The dyeing of weak acid dyes can be carried out in a weak acid bath or neutral bath. When dyeing in the weak acid bath, we generally control its pH value at 4 to 6. At this time, because it is near the isoelectric point of silk fibers, some dyes are still bound to silk fibers through ionic bonds. There are also some dyes that do not possess enough hydrogen ions in the dye solution to make the silk fibers have positive charges. Instead, it is dyed by hydrogen bonds and van der Waals molecular forces. When it is dyed neutral, the pH value should be about 6 to 7 at this time. At this time, the pH value of the dyeing bath is greater than the isoelectric point of silk fibers. Therefore, the combination of dye and silk fiber mainly depends on molecular collision, and then dyes by hydrogen bond and Van der Waals force.

3. 酸性染料染色的工艺条件/Dyeing Process Conditions of Acid Dyes

（1）染色温度。染浴温度视染料性能而异，对于某些染料，在低温时就具有良好的染色性能。对于分子结构比较复杂的弱酸性染料，在溶液中分子间聚集倾向较大，因此温度升高可以降低染料的聚集程度。同时，温度升高，可促使纤维膨化，增加染料分子的动能，因而提高上染速率，促使染料分子进入纤维闪部，达到染色均匀的目的。但是如果温度过高，织物经长时间沸染，会造成丝纤维损伤，影响染色成品光泽和质量，一般染色温度控制在95℃左右即可。低温染色时，利用纤维膨化剂及释酸剂促染（染料需经筛选），染色温度可降低至70～80℃，可减少织物表面损伤，提高成品的质量。

Dyeing temperature. The temperature of the dye bath depends on the dye properties. For some dyes, they have good dyeing properties at low temperatures. For weak acid dyes with complex molecular structures, their intermolecular aggregation tendency in solution is large, so the aggregation degree of dyes can be reduced when the temperature increases. At the same time, when the temperature increases, it can promote fiber expansion, and then increase the kinetic energy of dye molecules. Therefore, this can improve the rate of dyeing and promote dye molecules into the fiber flash, and finally achieve the purpose of dyeing uniformity. However, if the temperature is too high, long-time boiling dyeing will cause silk fiber damage, thus affecting the luster and quality of dyed finished products. Therefore, we generally control the dyeing temperature at about 95℃.

When dyeing at a low temperature, we can use a fiber bulking agent and acid-releasing agent to promote dyeing (dyes need to be screened), and the dyeing temperature can be reduced to 70 to 80℃. In this way, the damage on the fabric surface can be reduced and the quality of finished products can be improved.

（2）染色时间。纤维膨化，染料分子扩散进入纤维内部，这个过程需要一定时间。扩散速度较慢的染料更需要足够的扩散、渗透以及移染时间。但是，染色时间也不宜太长，一方面会使生产效率下降，另一方面织物长时间在高温下浸渍会影响丝织物的强力，故丝织物染色一般控制在60分钟左右。

Dyeing time. When the fiber swells, the dye molecules diffuse into the fiber and its process takes some time. The dye with a slower diffusion rate needs more time for diffusion, penetration, and dyeing. However, the dyeing time should not be too long. On the one hand, too long a time will reduce production efficiency. On the other hand, the fabric at high temperatures for a long time will affect the strength of the fabric. Therefore, the dyeing of silk fabrics is generally controlled at about 60 minutes.

（3）染浴pH。在酸浴中，丝纤维能吸收氢离子而呈阳电荷性，促使它和染料阴离子结合，因而酸是促染剂，弱酸性染料对丝纤维亲和力较高，在弱酸浴或中性浴中都能染色，但在中性浴染色时，有些染料上染较差。因此，必须根据不同染料来调节染浴的pH，从丝纤维特性及染料给色量说，以弱酸性浴染色为佳，上染快，得色深。致酸剂一般用醋酸或醋酸铵盐，以达到逐步释酸的目的。但对匀染性较差的染料，则以中性浴染色为佳，加酸会因上染速率加快而使染色不匀，一般采用的方法是先中性浴染色，再逐步加酸。

PH of dyeing bath. In an acid bath, silk fibers can absorb hydrogen ions and become positively charged. At this time, it can be combined with a dye anion. Therefore, acid is a dyeing promoter. Weak acid dyes have a high affinity for silk fibers and they can be dyed in a weak acid bath or neutral bath. However, some dyes have poor dye uptake in neutral bath dyeing. Therefore, we must adjust the pH of the dyeing bath according to different dyes. From the point of view of the characteristics of silk fibers and the amount of dye, we usually think that it is better to dye with a weak acid bath because of its fast speed of dyeing and deep color. We usually use acetic acid as the acidogenic agent or ammonium acetate. In this way, the gradual release of acid can be achieved. However, for dyeing departments with poor levelness, we should use neutral bath dyeing as the best. After adding acid, the dyeing rate will be accelerated and the dyeing will be uneven. Our general method is to dye in the neutral bath first, and then gradually add acid.

（4）对电解质的影响。酸性染料在酸性浴中染色时（等电点以下），染料以离子键与纤维结合，这时电解质如食盐、硫酸钠等加入能降低染料的上染率，因而它们是缓染剂。但是当弱酸性染料在弱酸性浴或中性浴中染色时，染料与纤维的结合主要是氢键和范德瓦耳斯力，加入电解质，可以降低纤维对染料间相互的电荷斥力，促使染料上染，因而是促染剂。为了防止上染过快造成染色不匀，在染色过程中，可分次加入电解质。

The influence on electrolytes. In acid bath dyeing, the ionic bonds of acid dyes (below the isoelectric point) will combine with fibers. At this time, the addition of electrolytes such as salt and sodium sulfate can reduce the dye uptake rate. Therefore, they are retarding agents. However, when weak acid dyes are dyed in the weak acid bath or neutral bath, the combination of dyes and fibers mainly comes from hydrogen bonds and Van der Waals forces. At this time, the addition of electrolytes can reduce the mutual charge repulsion between dyes. This can promote the dyeing of the dye, so it is a dyeing promoter. In order to prevent uneven dyeing caused by too fast dyeing, electrolytes should be added at different times during dyeing.

（5）染色浴比。浴比是指织物重量和染浴体积之比。浴比的大小，根据所用染色机械差异而不同，如卷染机浴比较小，为1：（3 ~ 5），绳状染色机稍大，约为1 ：50，而挂染及星形架染色浴比要大得多。一般来讲，浴比小，上染率高；浴比大，染色容易，得色均匀，但是，残留在架浴中的染料较多，上染率低。因而在实际生产上，大浴比染色往往都采用“连桶”，可以使残留在染浴中的染料得到充分利用。

Bath ratio of dyeing. The bath ratio refers to the ratio of the weight of fabric to the volume of the dyeing bath. The size of the bath ratio varies according to the dyeing machinery used. For example, the bath ratio of the jigger is small, which is about 1：(3 ~ 5). The rope dyeing machine is slightly larger, about 1 ：50, while the bath ratio of hanging dyeing and star rack dyeing is much larger. Generally, the smaller the bath ratio is, the higher the dyeing rate grows; the larger the bath ratio, the easier it is to get a uniform color. However, there are more dyes left in the rack bath, while the dye uptake rate is low. Therefore, in actual production, the larger bath than dyeing often uses an “even bucket” . This can make full use of the dye remaining in the dyeing bath.

以12103双绉用绳状机染色工艺为例（表4–5）。

Take the dyeing process of the rope machine for 12103 crepe de chine as an example（Table 4–5）.

工艺流程：

Process flow:

坯绸进桶→缝头→出水（冷一次）→染色（30 ~ 80℃升温50min，80℃染色40min）→洗水（冷水洗三次）→出桶→半成品检验→缝头→烘燥→轧固色液→成品

Blank into barrel → seaming → water outlet (cooling once) → dyeing (heating at 30℃ to 80℃ for 50 minutes and dyeing at 80℃ for 40 minutes) → water outlet (cooling for three times) → discharging from barrel → semi–finished product inspection → seaming → baking → rolling and fixing color liquid → finished product

表4-5　染色工艺处方
Prescription of Dyeing Process

染料酸碱性/The acidity and alkalinity of dyes	处方/Prescription
弱酸性染料/Weak acid dye	0.71g/L
中性染料/Neutral dye	0.8g/L
平平加O/Peregal O	0.15g/L
JFC（M型）渗透剂/JFC (M type) penetrant	0.2g/L
醋酸钠/Sodium acetate	0.6g/L
醋酸/Acetic acid	0.6mL/L

根据染浴泡沫情况及染色丝绸手感，决定是否加入消泡剂和柔软剂。

We should decide whether to add foam inhibitor and softening agent according to the situation of dyeing bath foam and dyeing silk feel.

二、活性染料染色 /Dyeing with Reactive Dyes

活性染料因含有能与纤维反应的活性基团，以共价键结合从而提高染料的色牢度，故被广泛应用于纤维素纤维的染色和印花。一些用于棉、羊毛和锦纶染色的活性染料，对蚕丝及其织物的染色效果良好。

Because reactive dyes contain active groups that can react with fibers, which are covalent bonds and can be combined, so as to improve the color fastness of dyes. Therefore, it is widely used in dyeing and printing of cellulose fibers. Some reactive dyes originally developed for cotton, wool and nylon have good dyeing effects on silk and fabrics.

1. 活性染料的结构 /The Structures of Reactive Dyes

活性染料由三部分（即母体、连接基和活性基团）组成。染料母体，其结构与酸性染料相似，如带磷酸根的偶氮结构、蒽醌结构、酞菁结构等。活性基团可以和染料母体牢固地结合在一起，又能与纤维发生反应而结合。下面介绍几种常见的活性染料。

Reactive dye is composed of three parts: matrix, bridging group and reactive group. The structure of the dye parent is similar to that of acid dyes, such as azo structure with phosphate, anthraquinone structure, and phthalocyanine structure. The active group can be firmly combined with the dyematrix, and can react with the fiber to combine. Here are some common reactive dyes.

（1）三聚氯氰型。该类活性染料又可分为二氯均三嗪型和一氯均三嗪型两类。

Cyanuric chloride type. This kind of reactive dye can be divided into two types, dichlorotriazine type and monochlorotriazine type.

三聚氯氰与染料母体结合后，还有两个活泼氯原子，称为二氯均三嗪型，其结构通式可用下式来表示（“D” 代表染料母体）：

After cyanuric chloride has combined with the dye matrix, there are two active chlorine atoms. It is called dichlorotriazine type. Its general structural formula can be represented by the following formula（“D” stands for dye matrix）:

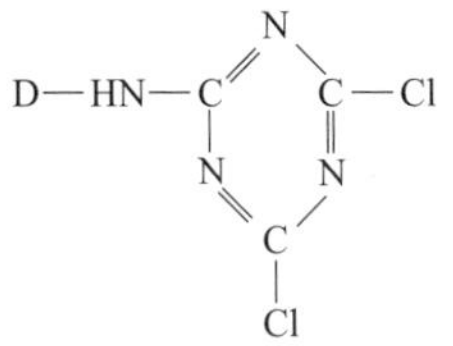

N
D—HN—C　C—Cl
N　N
C
Cl

如国产活性艳红 3B：

Such as domestic reactive brilliant red 3B:

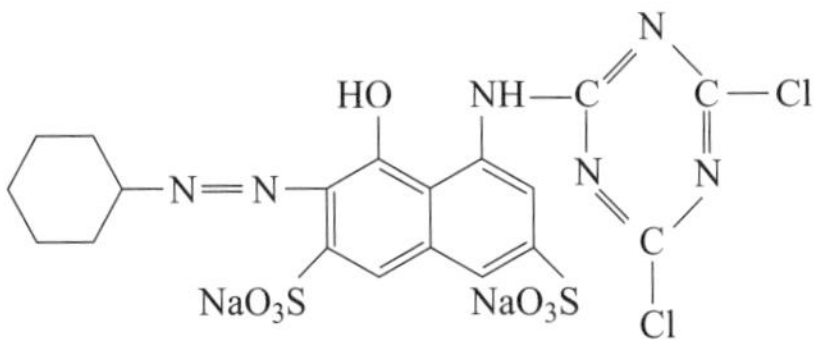

N
HO　NH—C　C—Cl
—N=N—
N　N
C
NaO_3S　NaO_3S　Cl

这类染料因为有两个活泼氯原子，化学性能活泼，能在较低温度及碱性条件下与纤维反应，但正因为它活泼，所以染料的稳定性较差，容易水解失活。

This kind of dye has two active chlorine atoms so its chemical properties are active. It can react with fiber under low temperatures and alkaline conditions. However, because of its activity, the stability of its dyes is poor, and it is easy to hydrolyze and then it would be inactivated.

三聚氯氰与染料结合后，两个活泼氯原子中一个被亚胺基取代，只剩下一个活泼氯原子，这类染料称为一氯均三嗪型。其结构通式为：

After cyanuric chloride has combined with the dye, one of the two active chlorine atoms is replaced by the imino group, leaving only one active chlorine atom. This kind of dye is called monochlorotriazine type. The general structural formula is:

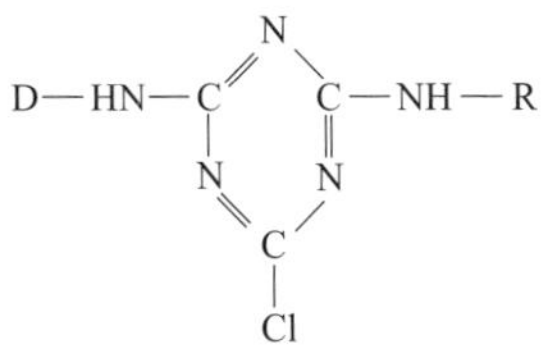

N
D—HN—C　C—NH—R
N　N
C
Cl

如活性黄 K–RN：

Such as active yellow K–RN:

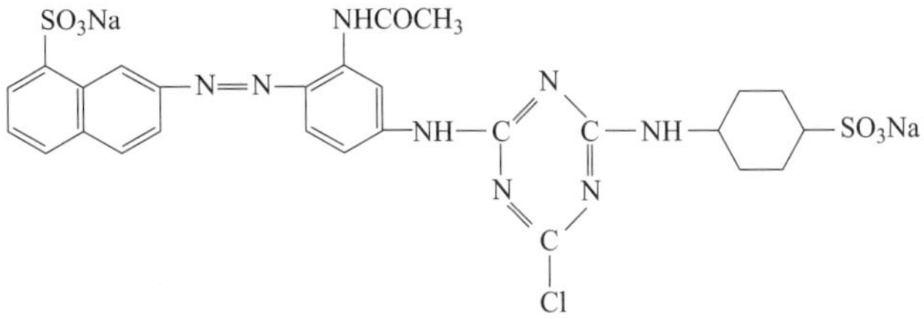

SO_3Na　$NHCOCH_3$
—N=N—　N
—NH—C　C—NH—　—SO_3Na
N　N
C
Cl

这类染料只有一个活泼氯原子，具有较高的稳定性，溶解时，加热无显著水解现象。染色时，得在较高温度和较高的碱性条件下才能与纤维反应结合，色牢度较高。

These dyes have only one active chlorine atom and are highly stable. If dissolving, it is heated without obvious hydrolysis. When dyeing, it needs a high temperature and high alkaline conditions to react with fiber, and its color fastness is high.

（2）砜型。这类染料的活性基团是乙烯砜（$—SO_2·CH=CH$）或 β－硫酸酯乙基砜（$—SO_2·CH_2·CH_2·OSO_3Na$）。国产活性染料KN型属于此类。

Sulfone type. The active groups of these dyes are divinyl sulfone ($—SO_2·CH=CH$)or β –sulphone–sulphatoethyl ($—SO_2·CH_2·CH_2·OSO_3Na$). The domestic reactive dye KN belongs to this category.

如国产活性艳红KN–5B：

Such as domestic reactive brilliant red KN–5B:

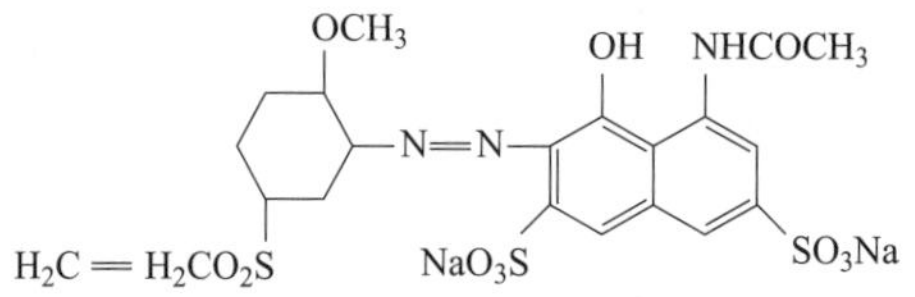

这类染料性质介于普通型和热固型染料之间，用碱量与普通型基本相同，但其固着温度比普通型的固着温度高，一般不低于60℃。

The properties of these dyes are between common and thermosetting types. The amount of alkali used is basically the same as that of the common type. However, its fixation temperature is higher than that of the common type. Generally, it is not less than 60℃.

（3）复合活性基团。此类染料具有两种不同的活性基团，提高染料的固着率，国产M型活性染料即属此类，它兼有一氯均三嗪和 β－硫酸酯乙基砜两种活性基团。

Compound active groups. This kind of dye has two different active groups and they can improve the fixation rate of dyes. M–type reactive dyes made in China belong to this category. It has two active groups: monochlorotriazine and β –sulphone–sulphatoethyl.

如国产活性艳红M–8B：

Such as domestic reactive brilliant red M–8B:

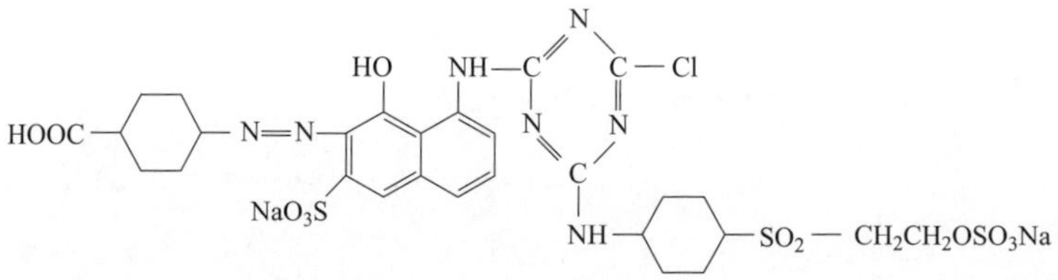

2. 活性染料对丝绸的上染机理 /The Dyeing Mechanism of Reactive Dyes on Silk

活性染料母体的化学结构基本与酸性染料基本相似，对丝纤维染色性能良好，色泽鲜艳，色牢度好。最佳染色条件因染料类型及染料母体的不同而不同，必须慎重选择。

The chemical structure of the reactive dye matrix is basically similar to that of acid dyes. It has good dyeing properties to silk fiber, bright color, and good color fastness. The optimum dyeing conditions are different with type and matrix, so we must carefully select them.

X 型活性染料在碱性、中性、酸性浴中均可对丝纤维染色。它在这三种体系中的上色机理有以下几种：

X–type reactive dyes can dye silk fibers in alkaline, neutral, and acid baths. Its coloring mechanism in these three systems can be described as follows:

（1）碱性染色。活性染料在纤维素纤维上的结合，主要借活性基团与纤维上的羟基发生反应而形成共价键。

Alkaline dyeing. The binding of reactive dyes on cellulose fibers mainly depends on the reaction of reactive groups with hydroxyl groups on the fibers to form covalent bonds.

其反应如下（“D”代表染料母体，“F”代表纤维）：

The reaction is as follows(“D” stands for dye matrix, “F” stands for fiber):

```
             N                                          N
           //  \              OH⁻                     //  \
D—NH—C        C—Cl +2F—OH ───→ D—NH—C        C—O—F
          |        ||                              |        ||
          N       N                                N       N
            \\   /                                   \\   /
              C                                        C
              |                                        |
              Cl                                       O—F
```

随着碱的浓度、温度和时间的不同，也可能发生如下反应：

Depending on the concentration, temperature, and time of alkali, this may also be as follows:

```
             N                                          N
           //  \              OH⁻                     //  \
D—NH—C        C—Cl+F—OH ───→ D—NH—C        C—O—F
          |        ||                              |        ||
          N       N                                N       N
            \\   /                                   \\   /
              C                                        C
              |                                        |
              Cl                                       OH
```

随着温度的提高（可能在染色的同时），有一部分染料水解：

With increasing temperature (and possibly at the same time as dyeing), a portion of the dye hydrolyzes:

```
             N                             N
           //  \       2OH⁻               //  \
D—NH—C        C—Cl ───→ D—NH—C        C—OH
          |        ||                 |        ||
          N       N                   N       N
            \\   /                      \\   /
              C                           C
              |                           |
              Cl                          OH
```

但在染色过程中，因为染料具有直接性，会很快地被纤维吸附，所以，上染的染料总要比水解的多。同时取决于工艺条件是否控制得当。

However, in the dyeing process, because the dye is direct, it will be quickly absorbed by the fiber. Therefore, there are always more dyes than hydrolyzed ones. At the same time, it depends on whether the process conditions are properly controlled.

对于活性染料如何和蚕丝纤维结合，有研究者认为，蚕丝纤维有能与活性染料起反应的基团，如氨基、亚胺基、羟基等，它们也能在碱性条件下与活性染料以共价键结合。

As to how to combine reactive dyes and silk fibers, some researchers believe that silk fibers also have reactive groups with reactive dyes, such as amino groups, imino groups, and hydroxyl groups. They also covalently bond with reactive dyes under alkaline conditions.

（2）酸性染色。在酸性浴中，活性染料也能很好地对蚕丝纤维染色，甚至能与酸性染料在同一浴中拼色，一般认为其染色机理与酸性染料相同，即在强酸性介质中以离子键结合，它的色牢度视染料母体的性能而异；在弱酸性浴中，染色则同时还要靠氢键和范德瓦耳斯力。在实践生产中，X型活性染料用酸性浴染色，往往染色牢度还达不到要求，通常还需固色。

Acid dyeing. In an acid bath, reactive dyes can also dye silk fibers well and even mix colors with acid dyes in the same bath. It is generally believed that its dyeing mechanism is the same as that of acid dyes, namely that it is in a strong acidic medium for ion bonding. Its color fastness depends on the properties of the dye matrix. Hydrogen bonding and van der Waals forces are needed to stain in a weak acid bath. In practical production, X–type reactive dyes are dyed in an acid bath. The color fastness is often not up to the requirements, so it is usually necessary to fix the color.

（3）中性染色。主要靠丝纤维对染料的吸附，借染料对纤维的亲和力上染，也就是说，与弱酸性染料在中性浴中上染原理相似。在生产实践中，活性染料在中性浴中染丝织物能得到较好的色牢度。

Neutral dyeing. It mainly depends on the adsorption of silk fibers to dyes and it dyes by virtue of the affinity of dyes to fibers. In other words, the dyeing principle of weak acid dyes in neutral baths is very similar to it. In production practice, the reactive dyes can get better color fastness by dyeing silk fabric in the neutral bath.

3. 活性染料的染色工艺/Dyeing Process of Reactive Dyes

活性染料在酸性浴中染色如弱酸性染料，一般采用醋酸调节至弱酸性，pH为4～6，此时，虽然丝纤维在等电点以上，带负电荷，与染料阴离子有排斥作用，但是染料对纤维的亲和力克服了这种排斥作用，而使丝纤维染色。加入中性盐，可以降低纤维表面电位，提高上染率，但中性盐加到一定程度后，上染率就逐渐下降，因此，要选择适当的用量及加盐的时间。在弱酸浴中，染色温度可适当提高，以提高上染率。一般来说，在80~90℃可获得最佳上染效果，但要根据染料的性能而定。

In acid bath dyeing, reactive dyes such as weak acid dyes, we generally use acetic acid to adjust to the weak acid, and the pH is controlled at about 4 to 6. At this time, although the silk fiber has a negative charge and repels the dye anion when the pH is above its isoelectric point, the affinity of the dye to the fiber overcomes this repulsive effect to dye up. After adding neutral salt, it can reduce the potential of the fiber surface and increase the dyeing rate. However, when neutral salt is added to a certain extent, the dyeing rate gradually decreases. Therefore, we should choose the appropriate amount of salt and the time of adding salt. In the weak acid bath, the dyeing temperature can be appropriately raised, thus increasing the dyeing rate. Generally, the best results are obtained at about 80℃ to 90℃, but it depends on the properties of the dye.

活性染料在中性浴（或弱酸性浴）中染色，在碱性浴中固着，色牢度较好。一方面，对于脱胶后的丝织物，如果在碱浴中长时间高温处理，容易引起损伤和茸毛；另一方面，X型活性染料在碱性浴中也容易水解，所以要选择合适的工艺条件，既不损伤丝纤维，又要提高上染固着率，达到最佳的染色效果。

Reactive dyes are dyed in the neutral bath (or weak acid bath), and fixed in the alkaline bath with good color fastness. For one thing, the silk fabric after degumming, if it is put in the alkali bath for a long time at a high temperature, it will easily cause damage and fuzz. For another, X-type reactive dyes are easily hydrolyzed in an alkaline bath. Therefore, we should choose an appropriate technological condition. In other words, it not only damages silk fibers, but also improves the dyeing fixation rate of the upper dyeing, so as to achieve the best dyeing effect.

以丝绸在无张力胶辊卷染机上染色为例，采用X型等活性染料在中性浴中的染色工艺：

Taking silk dyeing in jigger as an example, we use the neutral bath dyeing process of X-type and other reactive dyes:

织物润湿打卷→染色（40℃始染，以4道升温至90℃染4道后，沸染6道）→水洗（室温洗1~2道，80~100℃水洗2~4道，室温水洗1~3道）→冷水上卷

Wetting and rolling of fabric → Dyeing (starting at 40℃, after 4 passes of heating to 90℃ and 4 passes of dyeing, then boiling for 6 passes) → Washing (1 to 2 passes at room temperature, 2 to 4 passes at 80℃ to 100℃, 1 to 3 passes of washing at room temperature) → Rolling in cold water

染色处方见表4-6。

The prescription of dyeing is shown in Table 4-6.

表4-6　染色处方
Prescription of Dyeing

材料/Material	参数/Parameter
活性染料含量（对织物重）/Content of reactive dye (weight to fabric)	x%
食盐（或元明粉）/Salt (or sodium sulfate)	20～30 g/L
液量/Liquid volume	300L

采用中性染色法进行浸染时，宜在接近沸点（95℃）时，恒温染色30～50min，使染料有足够的能量与丝素反应，而不必再增加丝素的反应活性。

When the neutral dyeing method is used for dip dyeing, it is suitable to dye at a constant temperature for 30 to 50 minutes when the boiling point (95℃) is approaching. In this way, the dye can have enough energy to react with silk fibroin, so it is unnecessary to increase the reactivity of silk fibroin.

三、阳离子染料染色 /Dyeing with Cationic Dyes

阳离子型染料的发色基团大多数为有机碱类，能与无机酸结合形成盐，多数为盐酸盐，少数为其他盐类。阳离子染料在水溶液中离解时，色素基团带正电荷。

Most chromophore groups of cationic dyes are organic bases. It can form salts with inorganic acids. Most of them are hydrochloride salts, while a few are other salts. When cationic dyes dissociate in aqueous solution, the pigment groups are positively charged.

1. 阳离子染料的结构/The Structures of Cationic Dyes

阳离子染料的种类较多，主要介绍以下几类：

There are many kinds of cationic dyes, mainly including the following categories:

（1）三芳甲烷型。这种结构的染料占碱性染料的大多数，如盐基品红：

Triarylmethane type. Dyes with such structures account for the majority of the basic dyes, such as basic magenta:

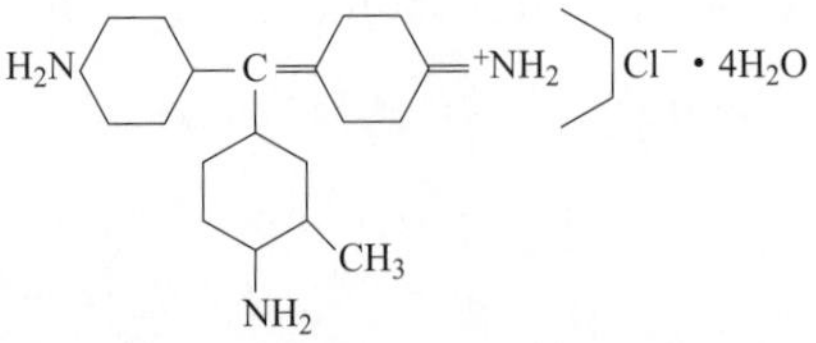

（2）偶氮型。如阳离子红2GL：

Azo type. Such as cationic red 2GL:

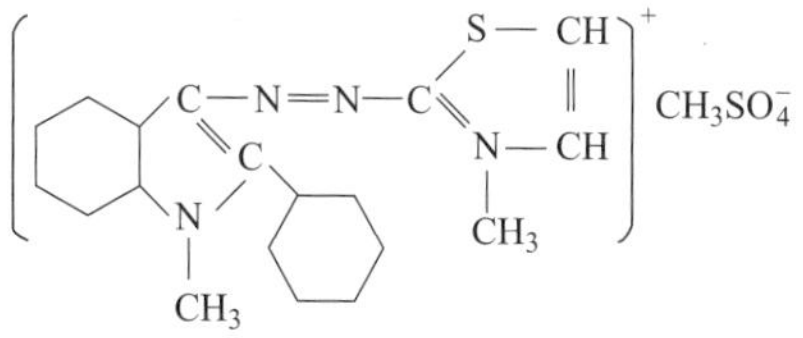

（3）噻唑类。如碱性亚甲基蓝BB：

Thiazole type. Such as alkaline methylene blue BB:

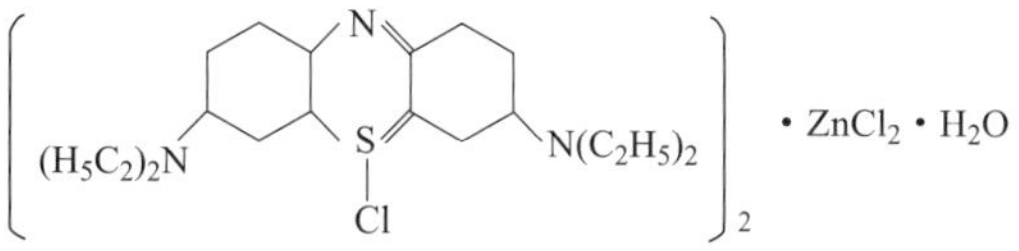

2. 阳离子染料对丝绸的上染机理/Dyeing Mechanism of Cationic Dyes on Silk

阳离子染料对丝绸的染色机理是丝纤维对阳离子染料有较大的亲和力。这个反应基本上是一个成盐过程。在染色的初期，染料阳离子迅速地被吸附在纤维表面上，这是由于织物带负电荷表面的静电吸引。在加热时，染料阳离子渗透进纤维内部，与纤维内部的酸性位置相结合。

The dyeing mechanism of cationic dyes on silk is that silk fibers have a great affinity for cationic dyes. The reaction is basically a process of salt formation. At the initial stage of dyeing, dye cations are rapidly adsorbed on the fiber surface due to electrostatic attraction on the negatively charged surface of the fabric. When heated, dye cations penetrate the interior of the fiber, which can be combined with the acid position inside the fiber.

3. 阳离子染料的染色方法/Dyeing Method of Cationic Dyes

阳离子染料上染丝织物的常规方法有以下三种：

There are three conventional methods for dyeing silk fabrics with cationic dyes:

（1）中性浴染色。为了避免染色不匀，可分多次加入阳离子染料。

Neutral bath dyeing. Aiming to avoid uneven dyeing, we can add cationic dyes in several times.

（2）碱性皂浴染色。如果利用橄榄油皂或工业皂染色，用量为织物重量的10%～15%，与染料一起染色。

Alkaline soap bath dyeing. If we use olive oil soap or industrial soap for dyeing, the amount is 10% to 15% of the weight of the fabric, and the dye is dyed together.

（3）酸性浴。用染料与醋酸一起染色，醋酸用量为织物重量的1%～2%。在50℃时开始染色，70～80℃染色1小时。染色后，用单宁酸和酒石酸处理，虽可提高湿牢度，但影响染色绸的光泽和手感。

Acid bath. When dyeing with dyes and acetic acid, the amount of acetic acid is 1% to 2% of the fabric weight. It was dyed at 50℃ and kept at 70 to 80℃ for one hour. After dyeing, we treated it with tannic acid and tartaric acid. Although this can improve its wet fastness, the luster and handle of dyed silk will be affected.

四、直接染料染色 /Dyeing with Direct Dyes

直接染料在丝织物上表现出良好的色牢度，所以它仍被作为丝绸印染染料。以弥补酸性染料色谱的不足，尤其在深色色谱方面，如深棕、墨绿、黑色等色泽。

On silk fabrics, the color fastness of direct dyes is also high. Therefore, it is still used as silk printing and dyeing dye. It can be used to complement the deficiency of acid dye chromatography, especially in dark chromatography, such as dark brown, dark green, black, and other colors.

1. 直接染料的化学结构 /Chemical Structures of Direct Dyes

直接染料分子一般都含有一个或多个磺酸基、羧基等可溶性基团的双偶氮、多偶氮结构。双偶氮染料多为黄色、红色、蓝色，而三偶氮、四偶氮染料则是绿色、深蓝、黑色等，这类偶氮染料，大致含有以下几种化学结构。

Direct dye molecules generally contain one or more Sulfonyl groups, carboxyl groups, and other soluble group of the double azo, polyazo structure. Double azo dyes are mostly yellow, red, and blue, while the triple and quadruple azo dyes are green, dark blue, and black. This kind of azo dye contains the following chemical structures.

（1）联苯胺结构。这类染料的制备都需用染料的中间体——联苯胺或其衍生物。近年来，有研究认为联苯胺有毒性，对人体有致癌作用，影响染料制造厂工人健康，因而联苯胺结构的染料（包括联苯胺制造的酸性染料），国内外染料厂都逐渐停产。

The structures of benzidine. The manufacture of these dyes requires the intermediate of the dye benzidine or its derivatives. In recent years, it has been considered that benzidine is toxic, which can cause cancer in the human body and affect the health of workers in dye manufacturers. Therefore, for benzidine structured dyes (including acid dyes made from benzidine), domestic and foreign dye manufacturers have gradually stopped production.

（2）具有其他共轭系统的偶氮染料。这类染料日晒牢度较好，一般在四级或以上，如直接耐晒黑G和FF，就是为了代替直接黑BN（联苯胺结构）而制造的。它的耐晒牢度可达六级，用以染人头发纤维织物乌黑度与BN黑接近，用以染丝织物，上染率和着色力均较低。

Azo dyes with other conjugated systems. Azo dyes with other conjugated systems. This kind of dye has a good light fastness, and it is generally at level 4 or above. For example, direct black G and FF are manufactured to replace direct black BN (benzidine structure). Its color fastness can reach level 6. It is used to dye the human hair with a blackness close to BN black, and it is used to dye the silk fabric with a low dyeing rate and coloring power.

直接耐晒蓝RGL：

Direct blue RGL:

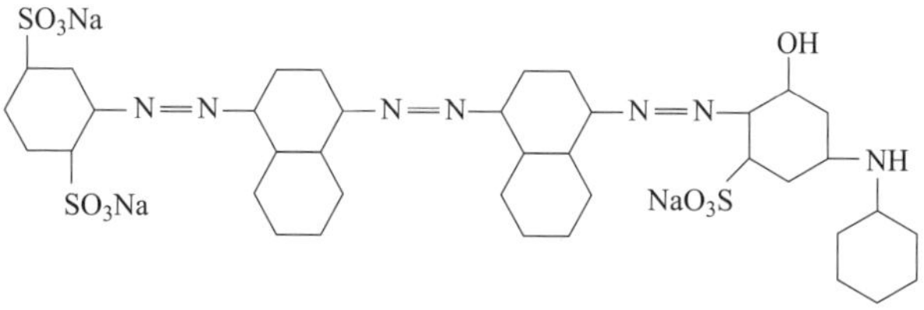

（3）具有尿素结构的偶氮染料。这类染料耐日晒性能都较良好，对纤维素纤维的直接性很强。在丝绸染色中，常用的有直接耐晒黄RS、红F3B、耐晒桃红G等。

Azo dyes with urea structure. This kind of dye has a good color fastness, and its directness to cellulose fiber is very strong. In silk dyeing, we commonly use such as direct yellow RS, direct red F3B, direct pink G, and so on.

直接耐晒黄RS：

Direct yellow RS:

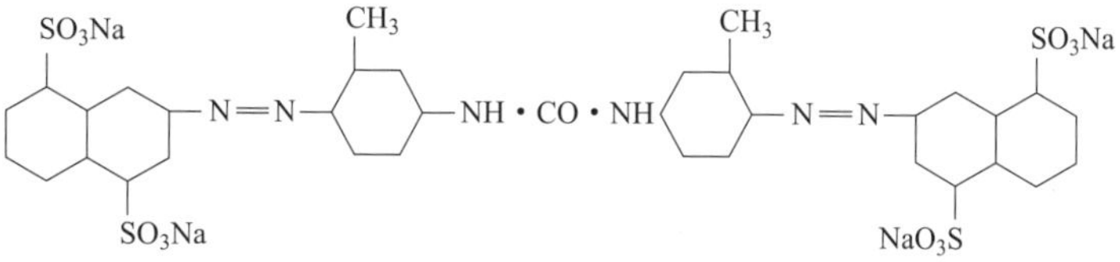

（4）具有三聚氰胺结构的偶氮染料。如直接耐晒绿5GLL，它是由三聚氯氰把偶氮结构的黄色组分和蒽醌结构的蓝色组分连接而成。

Azo dyes with melamine structure, such as direct green 5GLL, it is formed by connecting the yellow components of the azo structure and the blue components of the anthraquinone structure with cyanuric chloride.

直接耐晒绿5GLL：

Direct green 5GLL:

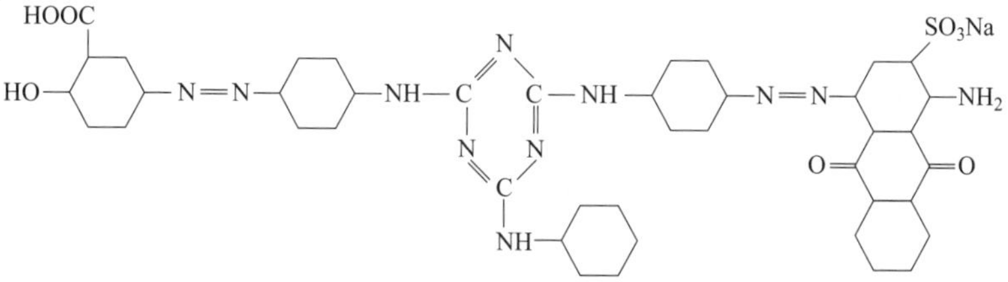

2. 直接染料对丝绸的上染机理/Dyeing Mechanism of Silk with Direct Dyes

直接染料分子中存在着羧基、羟基、氨基、偶氮基等基团，可与纤维形成氢键，同时直接染料分子呈线形和同平面性，以及有较长的共轭体系，有利于与纤维分子之间增加分子引力（范德瓦耳斯力），因此直接染料对纤维有较大的直接性。直接染料在蚕丝蛋白质上的染色原理与弱酸性染料在酸浴或中性浴中染色相似，它们在酸性介质中能形成离子结合，在中性浴中通过氢键和分子引力上染。

There are carboxyl groups, hydroxyl groups, amino groups, and azo groups in the direct dye molecules, which can form hydrogen bonds with the fiber. At the same time, the direct dye molecules are linear and coplanar, and have a longer conjugated system. It is beneficial to increase molecular forces (Van der Waals forces) with fiber molecules so that the direct dyes have a greater directness to the fiber. The dyeing principle of direct dyes on silk proteins is similar to that of weak acid dyes in an acid bath or neutral bath. They can form ion binding in an acidic medium, and dye in a neutral bath by hydrogen bonds and molecular forces.

3. 直接染料的染色工艺 /Dyeing Process of Direct Dyes

直接染料在丝织物上的应用大多采用深色及黑色。在中性染色时，有许多染料不能得到较高的上染率，因此必须选好染料。在酸性染色时，染浴需要控制在微酸性，这有利于纤维的物理性能和染料的上染，但要注意的是，上染不能太快。另外，在酸浴中产生沉淀的染料不能使用。

The application of direct dyes on silk fabrics is mostly dark and black. In neutral dyeing, there are many dyes that cannot get a high dyeing rate, so it is necessary to choose a good dye. In acid dyeing, the dye bath needs to be controlled to be slightly acidic. This is beneficial to the physical properties of the fiber and the dyeing of dyes. However, it should be noted that dyeing cannot be too fast. In addition, the dyes that precipitate in the acid bath should not be used.

直接染料在丝织物上的湿处理牢度视染料性能而异。如果想要得到良好的水洗牢度，一般需进行固色处理，固色剂及固色工艺同酸性染料。若成品还要经树脂整理，则可不必固色。

For the wet fastness of silk fabrics, direct dyes should be selected according to the different properties of dyes. However, if you need good washing fastness, then in general, you should need to fix the color. The fixing agent and fixing process used are the same as those of acid dyes. If the finished product needs to be finished with resin, it is unnecessary to fix the color.

五、天然染料染色 /Dyeing with Natural Dyes

由于合成染料迅速发展，天然染料逐渐被人们所忽视。近年来，人们发现，合成纤维和合成染料有引起某些人患皮肤病的可能。天然染料具有无毒、无害、无污染的优点。丝绸用天然染料染色有以下特点：色彩自然、优雅，兼有自然的香味，是合成染料所不能企及；染色的织物手感丰满厚实；用茜草、靛蓝、郁金香等染色的丝绸，具有防虫、杀菌作用，既利于保存，又能解决某些合成染料造成的过敏性问题。

Due to the rapid development of synthetic dyes, natural dyes have been neglected by people. In recent years, it has been found that synthetic fibers and dyes may cause some people to suffer from skin diseases. Natural dyes have the advantages of being nontoxic, harmless, and pollution-free. Dyeing silk with natural dyes has the following characteristics. Their colors are natural and exquisite, and they are beyond the reach of synthetic dyes with natural fragrances. The dyed fabric feels plump and thick. Silk dyed with rubia cordifolia, indigo, tulips, and so on, has insect control and bactericidal effects. It is not only good for preservation, but also for consumers who are allergic to some synthetic dyes.

1. 天然染料的分类 /Classification of Natural Dyes

按应用特点天然染料可分为以下几种：

According to their application characteristics, natural dyes can be classified as follows:

（1）天然色素，水溶性高，如黄檗能被纤维吸附，通常不需要添加媒染剂染色。

Natural pigment with high solubility in water. For example, phellodendron amurense can be adsorbed by fibers and it usually does not need mordant dyeing.

（2）天然色素几乎不溶于水，其配糖体可溶于水，可被纤维所吸附，需用后媒染法固着，如青茅草、栀子、槐、柏梅等。

Natural pigments that are almost insoluble in water. Its glycoconjugates are soluble in water and can be absorbed by fibers. At the same time, it is fixed by post-mordant dyeing. Such as a kind of grass called Qing Mao, cape jasmine, sophora, cypress, plum, and so on.

（3）天然色素在水中有相当良好的溶解度，可被纤维吸附。但为了提高染色牢度，需要使用媒染剂进行染色固定，如栌、苏枋（胭脂红或胭脂红酸）。

Natural pigment with fairly good solubility in water. It can be adsorbed by fibers. However, in order to improve the fastness, we have to dye it with mordants, such as a smoke tree, sappan wood (carmine or carminic acid), and so on.

（4）对水溶解度低的天然色素，有螯合物配位位置。因而与经先媒染而在纤维上吸附的金属离子形成配位健而固着，如紫草、茜素。

Natural pigments with low solubility to water. It has a chelate coordination position. Therefore, it forms a coordination bond with the metal ions adsorbed on the fiber through pre-mordant dyeings, such as the grass roots of a plant called Zi Cao and alizarin.

（5）植物染料中的天然色素在染色过程中，会在纤维上形成水不溶性的染料，如靛蓝、贝紫。

Natural pigments in plant dyes. It forms water-insoluble dyes on the fiber during dyeing, such as indigo and Beizi (a purple dye made from the gland of shellfish).

（6）利用对酸或碱溶解度的不同，使染料在纤维上吸附固着，如红花、郁金香。

The differences in its solubility in acid or alkali. We can make use of them so that the dye can be adsorbed and fixed on the fiber, such as safflower and tulip.

（7）丹宁的使用。可以使纤维吸附丹宁，主要经过媒染剂染色后固定颜色。

The uses of tannin. We can make the fiber adsorb tannin, which is mainly dyed by post-mordant and fixed in color.

从化学结构来看，大部分植物染料是复杂的酚类化合物。

In terms of chemical structure, most plant dyes are complex phenolic compounds.

2. 天然染料对丝绸的上染机理/Dyeing Mechanism of Natural Dyes on Silk

天然染料对丝绸和羊毛等蛋白质纤维能充分上染。为了提高色牢度，并呈现出多种色相，一般仍采用媒染法。茜素可以作为天然染料与媒染剂反应的一个例子。它是蒽醌系色素，与铝系媒染剂反应，结果形成络合物，使坚牢度改善。

Natural dyes can fully dye protein fibers such as silk and wool. In order to improve the colorfastness and show a variety of colors, we generally still use the mordant dyeing method. Alizarin can be used as an example of the reaction between natural dyes and mordants. It is an anthraquinone pigment. Its reaction with aluminum mordant results in the formation of the coordination complex, so as to improve the fastness.

3. 天然染料染色方法/Dyeing Method of Natural Dyes

（1）染液的制备。将树皮、树干、根、叶等植物切碎，放入有水的容器中，加热20～30分钟，提取煎汁。将此方法反复数次，提取煎汁，用以制备染液。

Preparation of dye liquor. We put tree bark, trunk, roots, leaves, and other plants in a container with water. Then heat it for 20 to 30 minutes, and we extract the decoction. This method is repeated several times and we extract the decoction, so that it can be used to prepare dye liquor.

（2）浸染。将上述染液稍加热，把丝线或丝绸浸入其中，沸煮10～20min。

Infiltration. We can heat the above dye liquor, then immerse the silk thread or silk in it, and boil it for 10 to 20 minutes at the same time.

（3）媒染。煮染后，放置染液直到冷却，即进行媒染。媒染剂根据不同色彩要求选择，可使用含铝、铜、铁或锡等化学品。将经过煮染的丝线和丝绸在含这些媒染剂的染液中浸渍媒染30分钟左右，然后水洗。

Mordant dyeing. After boiling and dyeing, we leave the dye solution until it cools, at which point we start dyeing. According to the different color requirements, we should choose different mordants. For example, chemicals containing aluminum, copper, iron, or tin can be used. We dip the boiled silk thread and silk in the mordant solution containing these dyes for about 30 minutes, then wash them with water.

以上为先染后媒染法。先媒染后染法则是先进行媒染，再染色。其顺序为：

The above is the method of first dyeing and post-mordant dyeing. The method of mordant dyeing first and then dyeing is mordant dyeing first, and then dyeing again. The order is as follows:

媒染→水洗→染色→水洗→干燥

Mordant dyeing → Washing → Dyeing → Washing → Drying

无论先染后媒染还是先媒染后染，都可根据需要进行一次或多次染色。就染色浓度而言，先媒染后染的方法染色深度高，但就匀染性来说，先染后媒染的方法好。

Whether dyeing before mordant dyeing or mordant dyeing first, it can be carried out once or repeatedly as needed. As far as dyeing concentration is concerned, the method of mordant dyeing before dyeing has a high depth. However, as far as levelness is concerned, the method of first dyeing and mordant dyeing is good.

第四节　印花/Printing

丝绸印花方法因设备而异。目前国内主要采用筛网印花设备，包括手工筛网印花台板、半自动筛网印花机、平版式自动筛网印花机、圆网筛网印花机等。圆网印花机是目前丝织物印花采用最多的设备。圆网印花最大的特点是花回尺寸较大，花型轮廓清晰，色彩浓艳，泥点细腻，立体感强，富于层次，适用于多套色大花回印花。

Silk printing methods vary from equipment to equipment. At present, screen printing equipment is mainly used in China, including the plate of manual screen printing, machine of semi-automatic screen printing, machine of flat screen printing and rotary screen printing, etc. The Printing machine of the rotary screen is the most widely used method for silk fabric printing at present. The biggest characteristic of screen printing is that the flower size is large and the pattern outline is clear. It is rich in color and has a good sense of three-dimensional strong, and rich in hierarchy. Therefore, it is suitable for printing with multiple sets of colors and large flowers.

丝绸印花按照工艺和方法可分为直接印花、拔染印花、防染印花、渗化印花、渗透印花等。

Silk printing can be divided into direct printing, discharge printing, anti-printing, infiltration printing, penetration printing, and so on.

一、直接印花/Direct Printing

直接印花，就是将色浆“直接”通过筛网花板刮印到织物上，经汽蒸固着后形成花纹。直接印花是织物印花的基本印花方式之一，可以与多种染料共印。

Direct printing means scraping and printing the color paste onto the fabric “directly” through the screen. After steaming and fixing, it forms many patterns. Direct printing is one of the basic printing methods of textile printing, which can be co-printed with various dyes.

1. 直接印花的工艺流程（图4-1）/Process Flow of Direct Printing (Figure 4-1)

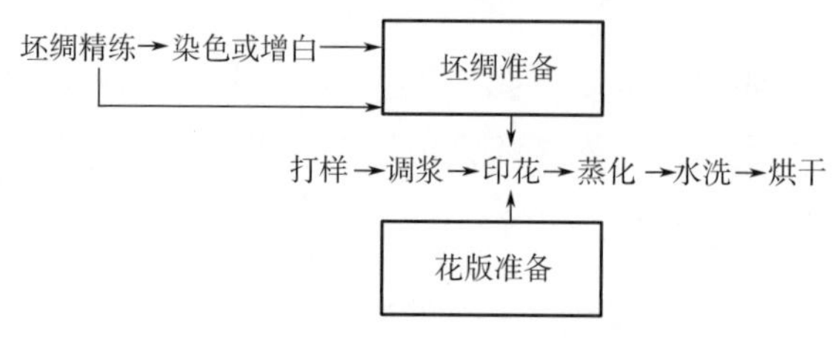

图4-1　直接印花工艺流程
Process Flow of Direct Printing

2. 直接印花色浆调制及处方举例/Direct Printing Color Paste Modulation and Prescription Examples

（1）11207真丝电力纺、19005斜纹绸使用小麦淀粉浆或白糊精淀粉混合浆为原糊；12107、12101等真丝双绉用可溶性淀粉浆为原糊。为使色浆易于洗涤，并增加成品花色鲜艳度，大多采用醚化植物种子胶或其混合糊。

11207 silk electricity texture and 19005 twill silk are made of wheat starch or dextrin to mix the paste as the original paste. 12107, 12101 and other silk crepe de Chine are made of soluble starch paste. In order to make its color paste easy to wash and to increase the brightness of its finished product, we mostly use etherified plant seed gum or its mixed paste.

直接印花色浆处方见表4-7。

Prescription of direct printing color paste is shown in Table 4-7.

表4-7　直接印花色浆处方
Prescription of Direct Printing Color Paste

处方/Prescription	用量/Dosage
原糊/Original paste	1000g
染料/Dye	*x*g
水/Water	200～250g
尿素/Urea	50g

（2）平板网印花机印色浆。19005斜纹绸、12107双绉、11207电力纺机印色浆以红泥三合浆为原糊（红泥浆15%，海藻乳化浆35%，海藻酸钠浆50%），其色浆处方如下（表4-8）。

Printing color paste by flat-screened printing-machine. 19005 twill silk, 12107 crepe de Chine, and 11207 electric textile machine printing pastes are made of red mud triple paste (15%

red mud, 35% emulsion paste of seaweed and 50% sodium alginate paste), and its prescription is as follows(Table 4–8).

表4–8 平板网印花机印色浆处方
Prescription of Printing Color Paste by Flat-screened Printing-machine

处方/Prescription	用量/Dosage
原糊/Original paste	1000g
染料/Dye	*x*g
水/Water	200～250g
尿素/Urea	70g
甘油/Glycerin	15g

（3）圆网机印色浆。丝缎印花采用小麦三合浆（淀粉浆50%，海藻浆17%，乳化浆33%）。圆网机印色浆处方如下（表4–9）。

Printing color paste by rotary screen machine. Silk and satin printing are made of wheat triad slurry (starch paste 50%, seaweed paste 17%, emulsion thickener 33%). Prescription of printing color paste by rotary screen machine is as follows (Table 4–9).

表4–9 圆网机印色浆处方
Prescription of Printing Color Paste by Rotary Screen Machine

处方/Prescription	用量/Dosage
原糊/Original paste	1000g
染料/Dye	*x*g
水/Water	250～300g
尿素/Urea	100g

二、拔染印花/Discharge Printing

在丝绸印花中，拔染一直受到重视。它能在深地色上得到浅花细茎的效果，使花纹具有鲜明的清晰度和立体感，但印花难度也相应增加。

In silk printing, discharge dyeing has been attached to great importance. It can get extremely detailed printing and dyeing effects in the case of deep color so that the pattern has a distinct clarity and three-dimensional sense. However, the printing difficulty is correspondingly increased.

雕白粉用作还原拔染，要在中性（拔白）或碱性介质（色拔及拔白）中进行，拔染色浆使用还原染料，其作用如下：

White powder (sodium formaldehyde sulfoxylate) is used for reductive discharge. It should be carried out in a neutral (white discharge) or alkaline medium (colored discharge and white discharge). Reductive dyes are used in the discharge dyeing size. Its functions are as follows:

$$NaHSO_2 \cdot CH_2O \cdot 2H_2O \xrightarrow{\triangle} NaHSO_2 + CH_2O + 2H_2O$$

$$NaHSO_2 + H_2O \longrightarrow NaHSO_3 + 2[H]$$

所产生的还原氢能使地色上的偶氮结构染料的偶氮键断裂，变成无色物质，使花纹处色泽消失。

The reduced hydrogen produced can make the azo bonds of azo structural dyes on the ground color disconnected and turn them into colorless substances, with the color of the pattern disappearing.

$$R_1-\langle\bigcirc\rangle-N=N-\langle\bigcirc\rangle-R_2 \xrightarrow{4[H]} R_1-\langle\bigcirc\rangle-NH_2 + R_2-\langle\bigcirc\rangle-NH_2$$

色拔浆采用还原染料，属于靛属、硫靛属或蒽醌结构染料。在碱性条件下，染料经汽蒸，被雕白粉还原溶解染着于纤维，然后氧化，在纤维上形成坚牢的色淀，反应大致如下：

Reductive dyes are applied for the size of the colored discharge, which belongs to indigo, thioindigo or anthraquinone dyes. Under alkaline conditions, the dye is steamed and reduced, dissolving onto the fiber by white powder, and then oxidizes again, with a firm color lake forming on the fiber. Its reaction is roughly as follows:

O, O, R $\xrightarrow[2NaOH]{2[H]}$ ONa, ONa, R $\xrightarrow{O_2}$ O, O, R

拔染印花处方见表4-10。

Prescription of discharge printing is shown in Table 4-10.

表4-10　拔染印花处方
Prescription of Discharge Printing

处方/Prescription	用量/Dosage
原糊（可溶性淀粉）/Original paste (soluble starch)	700g
雕白粉/White powder	80～150g
六偏磷酸钠/Sodium hexametaphosphate	5g
涂料白FTW/Paint white FTW	10g

续表

处方/Prescription	用量/Dosage
增白剂WS/Brightener WS	2g
水/Water	1000g

三、防染印花/Resist Printing

从工艺的角度来解释，防染印花是在织物上先印上防染浆料（色防或防白），再罩印其他色浆，不能改变其原印颜色的一种印花方法。在丝织物印花生产实践中，因使用的防染剂氯化亚锡也是拔染剂，它既能防也能拔，所以不管防印浆先印或后印，统称“防印”。防染印花方法实际上与直接印花相同。不同之处在于，氯化亚锡需要加入显示花纹的色浆中，因此，在刮印后产生一系列的化学变化。因为在防印过程中加入了氯化亚锡，可在工艺上作各种巧妙的变化，使丝绸印花的工艺效果更加丰富多彩。

From the technical point of view, resist printing is a printing method in resist printing paste (color or white anti-printing) which is printed on the fabric, and then other color pastes are overprinted. However, the original printing color cannot be changed. In the production practice of silk fabric printing, a resist agent (stannous chloride) is also a discharge agent, and it can resist and pull out. Thus, we collectively call it “resist printing” regardless of whether the anti-printing paste is printed first or after. The resist printing method is the same as direct printing. The difference is that stannous chloride needs to be added to the color paste showing the patterns. Therefore, it produces a series of chemical changes after scratch printing. Because of the stannous chloride added to the resist printing process, it can make various ingenious changes in the process. Therefore, it can make the technological effect of silk printing more colorful.

防染印花中还有一种半防法，其方法是先印一种含有匀染剂（如平平加O）的半防白浆，再罩印上部分色浆，那么色浆遇防白浆部分因其染料浓度被稀释和产生缓染作用，因此给色量降低，最后形成和原色调色光相同而色泽较淡的深浅两级。用这个方法，一只防白浆可和许多色浆相交，而得到许多颜色的深浅层次，从而达到少套多色的效果。

There is also a half-tone resistant printing method in the resist printing method. This method is to print a half-tone resistant whitening paste containing leveling agent (such as Peregal O) first, and then print some color paste. Then, the part of the color paste that meets the anti-whitening paste will be diluted due to its dye concentration and slow dyeing effect. Therefore, the color amount is reduced. And finally, two levels of light and dark, which are the same as the primary color toning light and lighter in color are formed. In this way, one resist-whitening paste can intersect with many color pastes to obtain many shades of color, so as to achieve a less set of multiple-color effects.

四、渗透印花/Penetration Printing

一般印花织物色泽、花纹均在正面制作，对织物反面不甚要求，但随着服装款式的发展，有的服装、方巾要求印花织物正反面色泽基本一致。渗透印花实为直接印花的一种，但因既要达到渗透效果，又要不影响色泽鲜艳度和印制轮廓清晰，对工艺则有更高的要求。

Generally, the colors and patterns of printed fabrics are mainly made on the front sides and their reverse sides are not required. However, with the development of clothing styles, some garments and kerchiefs require that the front and back colors of printed fabrics are the same. Penetration printing is a kind of direct printing. However, it is necessary to achieve the penetration effect, without affecting the brightness of color and clear printed outline, so we have higher requirements for the process.

影响丝织物印花渗透度的因素是多方面的，如织物的厚度、组织的规格、坯绸脱胶及杂质去除程度和匀净度，织物毛细管效应，染料、糊料、印花色浆的渗透性能，印制时刮刀的硬度、压力和刮印次数、贴绸用浆及贴绸方法以及印花台板温度等。

There are many factors that affect the printing penetration of silk fabrics. For example, the thickness, weave specification, removal degree and evenness of degumming and impurity of fabrics, as well as the penetrability of capillary effect, dyes, thickening agent and printing paste, and the hardness, pressure and scratch printing times, the paste and method for silk, as well as the temperature of the printing plate, etc.

渗透印花色浆处方见表4-11。

Examples of formulas for penetrating printing paste is shown in Table 4-11.

表4-11 渗透印花色浆处方
Prescription of Penetration Printing Paste

处方/Prescription	用量/Dosage
羧甲基纤维素/Carboxymethyl cellulose	1000g
酸性染料/Acid dye	*x*g
渗透剂/Penetrant	10～30g
尿素/Urea	60g
水/Water	250g
消泡剂/Foam inhibitor	2～3g

五、渗化印花/Infiltration Printing

渗化印花通过电子分色生产黑白印刷品，或采用人工分色描稿。采用适当的工艺路线和染化料及糊料，使印花绸富有立体感，出现由深到浅向四周渗化的印制效果。

Infiltration printing uses electronic color separation to produce black-and-white prints or manual color separation. The silk can be printed with a three-dimensional effect by the appropriate process, dyes, and thickening agents. The printing effect of bleeding from dark to light in all directions thus appears.

影响渗化印花的因素如下。

Factors affecting infiltration printing are as follows.

1. 织物组织 /Fabric Weave

一般织物越薄，渗化越容易；而织物越厚，组织越紧密，渗化越难。在渗化难易程度方面，双绉＞斜纹绸＞电力纺。

Generally, the thinner the fabric is, the easier the infiltration grows; the thicker the fabric is, the more compact its structure is and the more difficult the infiltration becomes. In terms of the difficulty of infiltration, that is crepe de Chine > twill silk > electricity texture.

2. 精练脱胶 /Scouring and Degumming

用于渗化印花的坯绸，毛细管效应宜在12cm以上，有利于染料向纤维内部扩散与渗化。

For the greige used for infiltration printing, the capillary effect should be more than 12 centimeters, facilitating the diffusion and infiltration of dyes into the fibers.

3. 感光制板 /Photosensitive Plate

采用电子分色直接将原样中各种颜色分别制成分色片，可加强由深到浅逐渐渗化的印制效果。同时，采用解像率高，在网孔中架桥性好，可制成精细、清晰图像的重氮感光胶进行感光制板，来提高渗化效果。

Electronic color separation is applied to directly make various colors in the original color into color separators, strengthening the printing effect from deep to shallow gradual infiltration. Besides, heavy nitrogen photoresist that a features high-resolution rate, good bridging in the mesh, and clear image is adopted for photosensitive plate, which can improve the effect of infiltration.

4. 糊料 /Thickening Agent

用于渗化印花的糊料，表面张力小、保水性差、含固量小、透网性好，因而宜将乳化糊与润湿性能好、保水性较差的可溶性淀粉浆以1：1的比例混合，再加入适量的渗透剂、扩散剂和吸湿剂等，并合理选用染料，以达到渗化印花的印制效果。

The surface tension of the thickening agent we used for infiltration is small. Its water retention is poor, the solid content small, and the mesh permeability good. Therefore, we should mix the emulsified paste with soluble starch slurry with good wettability and poor water retention in the ratio of 1：1. After that, we add a proper amount of penetrant, diffuser, hygroscopic agent, and so on. We need to choose dyes reasonably so as to achieve the printing effect of infiltration printing.

5. 染料/Dye

为了提高渗化效果，应选择扩散速率相差悬殊的染料拼色。就扩散速率来说，一般活性染料＞酸性、中性染料＞直接染料。在同色相配色时，选择扩散速率最快和最慢，色相一致的两种染料来拼色，经印花、蒸化、水洗，可在同色相中呈现出由深到浅的渗化效应。

In order to improve the infiltration effect, we should choose dyes with different diffusion rates. Generally, the diffusion rate of reactive dyes is higher than that of acidic dyes and that of neutral dyes is higher than that of direct dyes. In the same color matching, we choose the two dyes with the fastest and slowest diffusion rate and the same color. After printing, steaming, and washing, it can show from deep to shallow infiltration effect in the same color.

6. 助剂/Assistant

渗透、扩散、吸湿等助剂都影响渗化效果。就扩散性大小来说，NNO＞平平加O＞WA＞渗透剂YS＞拉开粉BX＞渗透剂JFC。在一些工厂，鉴于JFC有良好的稳定性，不会与染料发生沉淀，与纤维无亲和力等特点，而乳化浆中含有过量的平平加O，将两者拼用，有助于使染液形成深浅不同的层次，因而选用渗透剂JFC。此外，还在渗化印浆中加入太古油、尿素、甘油及古立辛A，以加强渗化扩散作用；加入硫酸铵，使印染浆料pH保持在6左右。

Assistant such as infiltration, diffusion, and moisture absorption will affect the infiltration. In terms of diffusivity, NNO > Peregal O>WA > Penetrant YS > Nekal BX > Penetrant JFC. However, in some factories, considering that JFC has good stability, it will not precipitate with dyes. It has no affinity characteristic with fibers and the emulsified slurry contains excessive Peregal O. When you combine the two, it helps to create different layers of dye. Therefore, penetrant JFC is selected by people. In addition, we can also add Turkey red oil, urea, glycerin, and thiodiglycol into the infiltration printing paste, to strengthen the infiltration and diffusion. At the same time, we need to add ammonium sulfate to keep the pH of the printing paste at about 6.

渗化浆基本处方见表4-12。

Basic prescription of slurry infiltration is shown in Table 4-12.

表4-12 渗化浆基本处方
Basic Prescription of Slurry Infiltration

处方/Prescription	用量/Dosage
乳化浆/Emulsion thickener	50g
可溶性淀粉浆/Soluble starch slurry	50g
太古油/Turkey red oil	15g
JFC	7g
古立辛/Athiodiglycol	5g

续表

处方/Prescription	用量/Dosage
尿素/Urea	10g
硫酸铵/Ammonium sulfate	3g
消泡剂/Foam inhibitor	2.5g

第五节 后整理/Post-finishing

蚕丝经精练、染色和印花诸多工序，最后进入后整理工序。后整理的主要目的如下。

After scouring, dyeing, and printing, silk finally enters the finishing process. The main purposes of post-finishing are as follows.

（1）改善织物的外观。丝绸通过整理，可产生柔和的光泽、优良的手感和悬垂性等特点。双绉还可具有特定的、均匀的绉效应。缎类织物还有丰满的手感，亮丽的光泽。

Improve the appearance of the fabric. Through finishing, silk can produce soft luster, excellent feeling, and draping characteristics. Crepe de Chine can also have a specific and uniform crepe effect. Satin fabrics also have a plump hand feel and a fat luster.

（2）改善织物的使用性能。丝绸经定形、拉幅、防缩整理后，具有规定的门幅和缩水率。服用时形状稳定。特别是经过汽蒸或呢毯整理机整理，可使织物在加工时受到影响的光泽和风格得到恢复。

Boost the service performance of the fabric. After setting, stretching, and anti-shrinkage finishing, the silk has the specified door width and shrinkage rate. Meanwhile, when in use, its shape is relatively stable. In particular, steam or cloth finishing machine, can restore the luster and style of the fabric affected during processing.

（3）增加织物的功能性、提高织物的附加值。通过各种化学整理，赋予蚕丝以抗泛黄、抗皱、防缩、增重、阻燃等新的性能；通过将化学整理与机械整理相结合，可改变织物的外观，使之具有褶皱和桃皮绒的外观，从而在不同程度上提高丝绸的附加价值。

Increase the functionality and added value of the fabric. Through various chemical finishing, it gives silk new properties such as anti-yellowing, crease resistance, anti-shrinkage, weight gain, and flame retardant. By combining chemical finishing with mechanical finishing, the appearance of the fabric can be changed, so that it has the appearance of wrinkles and peach skin velvet. Thus, it can improve the added value of silk to different degrees.

一、防泛黄整理/Anti-yellowing Finishing

丝绸老化是指蚕丝在日光、化学药品、湿度等因素的影响和作用下，强力显著下降和泛黄的现象。经过多年的研究，特别是由于已从老化泛黄的丝绸中部分地分离出了黄色酞分子和色原体。丝绸老化的原因主要有以下几种：①紫外线等光照作用，在紫外线等光照下，组成丝纤维的氨基酸，尤其是色氨酸、酪氨酸残基吸收光能，发生光氧化作用而变成有色物质，导致丝素强力下降。一般认为，酪氨酸在光照作用下可形成吲哚化合物，再变成有色物质。另外，色氨酸在光照下迅速分解，明显地呈黄褐色；②温湿度效应，丝织物在40%以上的湿度状态下放置，明显泛黄；③丝绸中杂质的影响，织物精练后残存的蜡质、有机物、无机物和色素等，以及练漂后未净洗的精练剂和漂白剂等，穿着时沾上的污垢，因洗涤不当而在织物上残留的洗涤剂，都可能引起丝绸泛黄、老化；④气体的作用，空气中的氧气，大气中的各种污染气体，如NO_x、SO_2等，以及塑料树脂加工过程中释放到大气中的各种污染气体，都能促进丝绸泛黄、老化。

The aging of silk refers to the obvious decrease in strength and yellowing caused by the influence and action of sunlight, chemicals, humidity, etc. After years of research, yellow phthalein molecules and chromogens have been partially separated from silk, especially aged and yellowed silk. The main reasons for silk aging are as follows: ①Ultraviolet light and other light effects. Under ultraviolet light, the amino acids, especially tryptophan and tyrosine residues, make up silk fibers absorb light energy and they become colored substances by photooxidation, which results in the decline of silk fibroin strength. It is generally believed that tyrosine can form indole compounds under light, and then it becomes colored substances. Tryptophan, on the other hand, breaks down quickly in light, and we can see that it has a distinct yellowish-brown color. ②Temperature and humidity effect. When the fabric is placed in a humidity state of more than 40%, it turns yellow obviously. ③The influence of impurities in silk.The residual wax, organic substances, inorganic substances, and pigment after scouring, and scouring agent and bleaching agent that have not been cleaned after scouring and bleaching, as well as the dirt when wearing, and detergent remaining on the fabric due to improper washing, they all may cause yellowing and aging of silk. ④The effect of the gas. Oxygen in the air, and various polluted gases in the atmosphere, such as NO_x, SO_2, and so on, as well as all kinds of polluted gases released into the atmosphere during the processing of plastic resin, can promote the yellowing and aging of silk.

丝织物防泛黄整理常用的化学试剂有：硫脲—甲醛树脂、含羟基化合物、肼化合物、环氧化合物、紫外线吸收剂、脱乙酰甲壳质、二元酸无水物等。下面列举丝织物防泛黄整理的实例：白色真丝电力纺常温下浸渍于含羟基的氨基甲酸酯树脂（树脂：水为1∶4和1∶6）、催化剂有机胺（树脂：催化剂为1∶0.5）溶液20分钟，离心脱水，使织物含液率为100%，60℃预烘20分钟，130℃热处理20分钟。树脂附着率分别为1.4%和2.6%。然后，在紫外线吸收剂（如2-羟基-4-正辛氧基二苯甲酮等）的1%溶液中浸渍2小时（浴比1∶50，常温，密闭），轻度脱液，烘干（30℃，24小时）。结果表明，由于发挥协同效应，防泛黄效果显著。

The chemical reagents commonly used for the anti-yellowing finishing of silk fabrics include thiourea-formaldehyde resin, hydroxyl compounds, hydrazine compounds, epoxy compounds, ultraviolet absorbent, chitosan, binary acid anhydrous substances, and so on.The following is an example of anti-yellowing finishing for silk fabrics. White silk of electric texture was immersed in the solution with hydroxyl-containing carbamate resin (the ratio of resin and water was 1 : 4 and 1 : 6) and catalyst organic amine (the ratio of resin and catalyst was 1 : 0.5) at room temperature for 20 minutes and then centrifuged to dehydrate so that the liquid content of the fabric is 100%. Next, it was pre-baked at 60℃ for 20 minutes and then heat-treated at 130℃ for 20 minutes. The adhesion rate of resin was 1.4% and 2.6% respectively. And then, it was immersed in a solution of 1% ultraviolet absorbent (such as 2-hydroxy-4-octyloxybenzophenone) for 2 hours (bath ratio 1 : 50, room temperature, closed). Next, it was slightly dewatered and then dried (30℃, 24 hours). The results show that this anti-yellowing effect is remarkable because of its synergistic effect.

二、抗皱整理/Crease Resistance Finishing

丝绸有易泛黄、起皱、易擦毛三大缺点。丝绸易起皱是由其本质所决定的。在不损害真丝绸原有优良特性的前提下，通过化学改性赋予其抗皱性，是抗皱整理的关键。丝绸的抗皱整理主要有以下三种方法。

Silk has three disadvantages: yellowing, wrinkling, and rubbing. Silk wrinkling is determined by its nature. Under the conditions of not damaging the original excellent characteristics of real silk, it is the key to give its crease resistant by chemical modification. There are three main methods for crease-resistant finishing of silk.

（1）用高分子树脂整理剂嵌入蚕丝纤维内部的间隙，并加以填充。通过赋予纤维弹性和蓬松性，从而使丝绸具有抗皱性。

We can use polymer resin finishing agents to embed the gap inside silk fibers and fill it. By giving fibers with elasticity and bulkiness, silk will have crease resistance.

（2）用化学整理剂使纤维分子间交联结合，形成网状的化学结构，防止纤维分子链的滑动，以达到抗皱的目的。

We can use chemical finishing agents to cross-link the fibers. It forms a reticular chemical structure, which can prevent the sliding of fiber molecular chains and finally achieve crease resistance.

（3）将上述两种方法结合起来，即填充树脂与交联整理剂结合使用。借两者的相乘效果，赋予丝绸以抗皱性。

We can combine the above two methods, namely that the filling resin is combined with the crosslinking finishing agent. By the multiplication effect of the two, we finally give silk with crease resistance.

纤维反应型树脂即*N*–羟甲基化合物作为丝绸抗皱整理剂是有效的，它能改善丝绸的抗皱性和防缩性，但有游离甲醛的问题。甲醛具有强烈的臭味，还会导致皮肤过敏症，使耐磨性下降。

As the finishing agent of crease resistance for silk, fiber reactive resin, namely *N*–hydroxymethyl groups, is effective. It can improve the crease resistance and shrink resistance of silk, but it has the problem of free formaldehyde. Formaldehyde has a strong odor and it can also cause dermatitis allergy, thus reducing the wear resistance.

通常，丝绸用合成树脂整理的方法是：丝绸浸泡在树脂液中，60 ~ 90℃烘燥 5 ~ 10 分钟，然后在 120 ~ 130℃焙烘 5 ~ 10 分钟，使树脂溶液以三维网状构造的聚合物与纤维中的活性基反应，形成分子链间的交联结合。最后，可用皂洗或碱洗，以去除纤维内残留的未缩合树脂和分解产物等。

Usually, the method of finishing silk with synthetic resin is: the silk is immersed in the solution of resin under the condition of 60℃ to 90℃ for 5 to 10 minutes, and then baked in the 120℃ to 130℃ for 5 to 10 minutes. In this way, the resin solution which reacts with the active groups in the fiber in a three–dimensional network structure polymer can form the cross–linking combination between molecular chains. Finally, we can use soaping or alkali washing to remove the residual uncondensed resin and decomposition products in the fiber.

三、防缩整理 /Anti–shrinking Finishing

丝绸是由经丝和纬丝交织而成的，经丝与纬丝形成波浪形的弯曲。在织造和染整加工过程中，在一定的张力和拉伸作用下，丝纤维经纬弯曲度减小。干燥时虽然暂时稳定，但一旦落入水中，在没有张力的状态下，丝纤维膨胀变粗，经纬丝就恢复到原来弯曲的状态而造成缩水。此外，蚕丝纤维具有良好的吸湿性。在湿态时，水分子易钻入丝纤维的空隙，使丝纤维溶胀，直径增加28.7%左右，而长度溶胀仅为1.7%，导致织物长度缩短，织造收缩率增加，缩水现象更严重。况且，丝绸经向丝线间隔较密，空隙较小，致经丝受湿热后膨化的余地较小，从而纬丝弯曲的机会也减少，这也是丝绸经向缩水率大而纬向缩水率小的原因。

Silk is made by interweaving warp and weft threads which can form wavy bends. In the process of weaving and dyeing and finishing, the warp and weft curvature of silk fiber decreases under certain tension and stretching. Although it is temporarily stable when it is dry, once it falls into the water, the warp and weft yarns will return to their original bent state and then cause shrink under the state of no tension. In addition, silk fiber has a good hygroscopicity. In the wet state, water molecules are easy to drill into the gaps of silk fibers, which makes silk fibers swell especially and its diameter will increase by about 28.7%. However, its swelling length is only 1.7%, which leads to the shortening of the fabric length and the increase of its weaving shrinkage, resulting in a more serious shrinkage phenomenon. Moreover, the warp threads of silk are closely spaced, and their gaps are small. The warp yarn is less likely to swell when exposed to moisture and heat conditions, thus reducing the chance of bending the weft yarn. This is the reason why silk has a large shrinkage in the warp and a small shrinkage in the weft direction.

经化学整理后再适当辅以机械整理，可提高丝绸的抗皱性。如12107双绉，原经、纬向缩水率分别为5.7%和1.7%，在精练、印花后，经整理浴一浸一轧，轧液率90%。首先，选用上海树脂厂有机硅羟乳SAH-288B 20g/L，SAH-283 5g/L、SAH-2（30%）5g/L、$MgCl_2 \cdot 6H_2O$ 2g/L、JFC 1 g/L、抗静电剂SN 5 g/L、Perlitsi-SW1 g/L。接下来，pH控制在6.5～7，然后在针板超喂，120℃烘焙30分钟，再呢毯整理。结果，除抗皱性显著改善外，经纬向的缩水率分别降为-0.1%和0.7%。

It is chemically finished and then properly mechanically finished, which can improve the cease resistance of silk. Take 12107 double crepes for example. It originally has a water shrinkage of 5.7% and 1.7% on warp and weft respectively. After refining and printing, it goes through the following finishing bath, one-immersing, and one-rolling, with the pick-up ratio reaching 90%. Firstly, we use PolysiloxaneSAH-288B 20 g/L from Shanghai Resin Factory, SAH-283 5 g/L, SAH-2(30%) 5 g/L, $MgCl_2 \cdot 6H_2O$ 2 g/L, JFC 1 g/L, antistatic agent SN 5 g/L, Perlitsi-SW 1 g/L. Next, we control its pH at 6.5 to 7 and then adjust the speed of its feed when shaping or others. Placed at 120°C and baked for 30min, it then enters the blanket finishing. As a result, the shrinkage in warp and weft directions decreases to -0.1% and 0.7%, respectively, in addition to the significant improvement of wrinkle resistance.

第五章 丝绸类非物质文化遗产
Intangible Cultural Heritage of Silk

中国丝绸不仅历史悠久，而且品种繁多，一些具有地方风格的代表性品种如雨后春笋般涌现，其中有苏绣、缂丝、宋锦、旗袍等。在吸收外来文化的同时，产品表现出东西方文化交融的时代特色，有的已经入选我国非物质文化遗产代表作名录。

Chinese silk not only has a long history, but also a wide variety, with some representative varieties with local styles springing up, among which there are Suzhou embroidery, cut silk brocade (kesi), Song brocade, cheongsam (Chinese dress), and so on. Assimilating the cultures abroad, the Chinese have produced numerous silk products that show the characteristics of cultural exchanges between the East and the West, some of which were inscribed on the representative list of Chinese Intangible Cultural Heritage.

第一节 苏绣/Su Embroidery

苏绣是苏州地区刺绣产品的总称，是江苏省苏州市的一种民间传统美术。苏绣起源于苏州，是四大名绣之一，是国家级非物质文化遗产之一。苏绣以针法精细、色彩雅致而著称。苏绣图案秀丽，题材广泛，技法活泼灵动。无论是人物还是山水，都体现了江南水乡细腻绵长的文化内涵。

Suzhou embroidery, Su embroidery for short, the general name of Suzhou embroidery products, is a traditional folk art in Suzhou, Jiangsu Province. Well known for its delicate patterns, sophisticated techniques, and a wide range of themes, it is one of the four famous embroideries, Su, Xiang (Guangxi), Shu (Sichuan) and Yue (Guangdong) and was inscribed on the representative list of Chinese intangible cultural heritage in 2006. Among the embroideries are portrayals of figures and landscapes, mirroring the cultural connotations of delicacy and constancy in the history of the south of the Yangtze River.

苏绣的发源地在苏州吴县一带，现已遍衍江苏省的无锡、常州、扬州、宿迁、东台等地。江苏省土地肥沃，气候温和，蚕桑发达，盛产丝绸，自古以来就是锦绣之乡。优越的地理环境，绚丽丰富的锦缎，五光十色的花线，为苏绣发展创造了有利条件。据西汉刘向《说苑》记载，早在两千多年前的春秋时期，吴国已将苏绣应用于服饰。宋朝以后苏绣的艺术水平达到了鼎盛时期。当时苏州是中国的锦绣之乡：绚丽多姿的锦缎，五光十色的花线，为苏绣的发展创造了有利条件。从而形成了苏州“户户有刺绣，家家有绣娘”的地方特色，在一方锦帛上可绣出五岳、江海、城邑、行阵等图案，并有“绣万国于一锦”之说。

Originating in Wuxian County, Suzhou, the embroidery has now gained popularity around Suzhou, such as Wuxi, Changzhou, Yangzhou, Suqian and Dongtai in Jiangsu Province. Home of embroidery since ancient times, Jiangsu has developed an industry of mulberry silkworm and silk weaving thanks to its fertile farmlands and gentle climate. Favorable conditions have been created for the development of Suzhou embroidery, such as a desirable geographical environment, rich brocade and satin, and colorful silk threads. According to *Garden of Stories*, written by Liu Xiang of the Western Han Dynasty, as early as the Spring and Autumn period more than 2,000 years ago, Su embroidery was used for clothing in Wu, an ancient state in this period. After the Song Dynasty, its artistic level reached its peak. Accordingly, there emerged local characteristics of “embroidery for every household and embroiderer for every family” in Suzhou. Some patterns were usually embroidered, such as five great mountains, rivers and seas, cities and countries, and combat forces, which was interpreted as “embroidery of all countries in a piece of brocade” .

苏绣以“精细雅洁”而闻名，在清朝进入了鼎盛时期，当时的苏州有“绣市”的誉称。清朝中后期，苏绣在绣制技术上有了进一步发展，新出现了精美的“双面绣”，仅苏州一地专门经营刺绣的商家就有65家之多。此时，苏绣诞生了沈寿、钱慧、曹墨琴等苏绣艺术大师，他们的作品往往价值连城。沈寿，原名沈云芝，字雪宧，苏州人，享有“绣圣”“神针”的美称，在传统刺绣的基础上，吸收了西洋油画在用光、用色和明暗关系上的技巧，改变传统刺绣中平铺直套的针法，根据物象的结构旋转用针，达到转折自然、细腻平服、光彩柔和的艺术效果；并将多种色线合一股施以滚针，随着丝理的转折变化互为映衬，完美表现了物象的明暗层次和立体感，达到了生动逼真的艺术效果，由此创造了享有“绣中之绣”美誉的仿真绣，把传统刺绣工艺推向新的高度。她在苏绣的基础上，借鉴西方油画的用光用色技巧，融合素描结构和阴阳向背原理，发明了能表现写实题材的针法，使绣品具有色彩明暗，从而达到了立体逼真的效果，图5-1是沈寿的作品《海棠玉兰》。此外，沈绣还创造了“拼色法”，也就是在绣制时把几种色线合为一线，穿到一个针孔里。这种方法类似于绘画创作的调色，能够并置得到新色，又能保留原有颜色，使画面色彩丰富多变，图5-2所示为沈寿作品《百合缠枝牡丹》。

Known for its exquisite elegance, Su embroidery entered its heyday in the Qing Dynasty, when Suzhou gained even more reputation as an embroidery city, where there were 65 embroidery shops. In the middle and late Qing Dynasty, it was further developed in embroidery technology, and a beautiful double-sided embroidery emerged. The city boasted a number of masters such as Shen Shou, Qian Hui and Cao Moqin, whose broideries were of great value.Shen Shou, formerly known as Shen Yunzhi, alias Xuehuan and a native of Suzhou, earned her reputation as an "embroidery saint" and "divine needle" . On the basis of traditional embroidery, she assimilates the skills of Western oil painting in the relationship between light and color, and light and shade, changing from the traditional needling method to rotating needling according to the structure of the patterns. This method helps to achieve the artistic effect of smooth transition, delicate texture, pleasant color, coupled with rolling needling and a variety of colorful silk threads, reflecting the contrast between light and dark and the three-dimensional effect of the patterns with the turning changes of silk texture, and thus creating an embroidery masterpiece, a step forward on traditional embroidery techniques. With the sketch technique and the principle of light contrast, she invented a needle method that can display realistic themes, as shown in *Begonia and Yulan Magnolia* (Figure 5-1), whose embroideries were later known as Shen Embroidery. She also created a color-matching method; that is, when embroidered, several color threads are twisted into one that passes through a pinhole.This method is similar to the tone of painting creation, which can be juxtaposed with new colors, but can also preserve the original colors, making the images rich and variable, as seen in *Lily Wound onto Peony* (Figure 5-2) .

图5-1　沈寿作品《海棠玉兰》
Shen Shou's Masterpiece *Begonia and Yulan Magnolia*

图5-2　沈寿作品《百合缠枝牡丹》
Shen Shou's Masterpiece *Lily Wound onto Peony*

民国时期，苏绣业一度衰落。中华人民共和国成立后，苏绣得到进一步的恢复和发展。

During the Republic of China, the Su embroidery industry once declined while it has been further restored and developed since the founding of the People's Republic of China.

在中国传统刺绣工艺中，苏绣擅长吸取其他艺术精华的品种，它的图案以亭台楼阁、小桥流水为题材，一般以蓝、绿为主色调，体现清雅、幽静的效果。从人物、花鸟到山水、动物，从静若处子到动如脱兔，苏绣呈现的是江南水乡那细腻绵长的人文内涵，如图5-3所示。苏绣作品的主要艺术特点是山水能分远近之趣、楼阁具现深邃之体、人物能有瞻眺生动之情、花鸟能报绰约亲昵之态。苏绣的仿画绣、写真绣其逼真的艺术效果是名满天下的。而这些美轮美奂的苏绣都是在上千年的历史时空中，一代代绣娘巧手穿引，心手相传，一针一线创造出来的。工艺性和艺术性的完美结合构成了一件好的苏绣艺术品，作品凝结的艺术效果也就成为鉴别苏绣工艺品和艺术品的重要标准。

Among the traditional Chinese embroidery craftsmanship, Su embroidery is good at learning the essence of other arts, which bases its design on a waterscape that covers pavilions, bridges, and waters. Generally, the main colors of the design, blue and green, show elegance and tranquility. It presents delicacy and constancy of the water township of Jiangnan, regions south of the Yangtze River, with its vivid patterns ranging from static to dynamic, as shown in Figure 5-3. Statically, the patterns of landscape and pavilions feature a vivid depiction of distance, flowers pleasance; dynamically, the figures eye's expressions, and birds affinity. This is why the embroidery is best known in China.The Su embroiderers created these beautiful crafts stitch by stitch, passing on the time-honored craftsmanship from one generation to another over thousands of years, a perfect combination of craftsmanship and artistry, and an important criterion for the identification of its artistic effect.

图5-3　苏绣作品
Su Embroidery Crafts

第二节　缂丝/Kossu

缂丝又称为刻丝（也写作克丝），是中国传统丝绸艺术的精华，是中国丝织业中最传统的一种挑经显纬，是非常具有欣赏和装饰价值丝织品。它的纹样边饰清晰，历历如刀镂一般。因而，织品的图案具有立体感，仿佛镶嵌在绸面上。

The essence of traditional Chinese silk art, kossu, also known as cut silk brocade, is one of the most traditional Chinese silk weaving techniques and a type of silk textile with appreciative and decorative value. The pattern of kossu, clearly trimmed like a knife, has a three-dimensional effect as if inlaid on silk.

图5-4　缂丝机
Kossu Machine

缂丝是一种特殊的丝织品，它和一般的丝绸织物不同。一般的丝绸织物都是经纬线通贯到底，不管是经线起花织物，还是纬线起花织物，织制方法都是相同的。而缂丝是经线贯通，纬线弯曲。这种织法可根据花纹图样，不受限制地将花纹在织物上织制出来在有花纹的地方进行缂织。图5-4所示为缂丝机。

Kossu is a special kind of silk fabric different from other silk fabrics. Generally, when silk fabrics are woven, warp and weft run through to the end. It is the same with two techniques: interwoven with warp or weft only where the pattern is desired. However, kossu is woven in a different way in which warp runs through while weft is interwoven with warp only where the pattern is desired, a weave that allows patterns to be woven on the fabric as expected. Figure 5–4 shows the kossu machine.

在唐朝，缂丝主要制作丝带等小件实用品，工艺以平纹编织为主，花地之间的交接处有明显的缝隙。纹样以简单的几何纹为主，色彩层次也不够丰富，主要是平涂的块面，还没有使用晕色。

During the Tang Dynasty, kossu were some small articles such as ribbons, and the technique was mainly plain weave, with obvious gaps available. The theme of pattern design is based simply on geometry, without multiple colors and halo.

在宋朝，精美的缂丝成为著名的丝织艺术品，达到了顶峰，缂丝也是从此正式定名，尤以大幅的绿丝画为特色，其品格高雅，多为宫廷所用，价格昂贵，古人以“一寸缂丝一寸金”之说。北宋的缂丝前承唐朝，但花纹更为精细富丽，纹样结构对称，富于变化，从而创造了“结”的戗色技法。无论包首、装裱等饰品，还是山水、花鸟、人物等缂丝艺术品，已达到相当水平。到了南宋时期，缂丝生产规模扩大，北方的缂丝技艺传入南方，使江南的缂丝工业超过了北方。镇江、松江、苏州的缂丝工业拔地而起，涌现了一批著名的名工巧匠，如松江的朱克柔，吴郡的沈子蕃、吴煦，还有朱良栋等名家。他们专门仿制唐宋名画家的书画，使缂丝这种丝织艺术向独立欣赏性的方向发展，开创了缂丝艺术的全新时代。

The exquisite kossu became a famous silk art in the Song Dynasty, when it gained wider popularity and was formally named as kossu presented as large green silk tapestries imitating paintings. It was valued in the royal palace, spiritually and materially. Kossu in the Northern Song Dynasty found its origin in the Tang Dynasty, but the patterns were more elaborate, symmetrical in structure, and rich in change, thus creating a “knot” dyeing technique. The craftsmanship of kossu reached a considerable high level, as found in such accessories as the head of bags, mounting, and in artworks such as landscapes, flowers, birds, and characters. The production of kossu in the Southern Song Dynasty was expanded, and the techniques of kossu in the north passed to the south, which made kossu industry in the south of the Yangtze River exceed that in the north. Kossu industry in Zhenjiang, Songjiang, and Suzhou emerged, together with a number of famous craftsmen such as Zhu Kerou in Songjiang, Shen Zibo and Wu Yan in Wujun, Zhu Liangdong, and other famous craftsmen. They epecially imitated the calligraphy and paintings of famous painters in the Tang and Song Dynasties, which made the silk weaving art of kossu develop aesthetically, and created a new era of art.

元朝，缂丝艺术被广泛应用于寺庙用品和官服上，并开始采用金彩。缂丝简练豪放，一反南宋细腻柔美之风，这对明清两代的缂丝艺术产生了很大的影响。

In the Yuan Dynasty, the art of kossu was widely used in temple supplies and official clothes and with golden threads. Kossu was then plain and bold, contrary to the delicate and gentle style of the Southern Song Dynasty, which had a great impact on silk art in the Ming and Qing Dynasties.

在明清时期，缂丝技艺更加精湛。缂丝作品除欣赏性的书画外，还有服装、台毯、围幔、坐垫、屏风、香荷包、官扇等各种生活必需品。在明朝，缂丝被誉为宫廷艺术，被皇家用来制作龙袍（图5-5）。时至晚清，随着国势衰弱，缂丝工业出现了濒临绝种的状态，缂丝粗劣之作充斥于市，即便宫廷所用之物也罕有精品。

The kossu technique became more sophisticated in the Ming and Qing Dynasties. Among the kossu products are a variety of life necessities such as clothing, table carpets, fencing, cushions, screens, incense bags, and official fans, besides calligraphy and painting. In the Ming Dynasty, kossu, known as the art of the court, was used by the royal family to make dragon robes (Figure 5-5). Even in the late Qing Dynasty, with the weakening of the country, the kossu industry was on the verge of extinction, when exquisite kossu products were rarely found even in the imperial court, with those of low-quality abundant in the market.

图5-5　明朝万历皇帝的“衮服”
Gunfu (emperors' dress) in the Ming Wanli Period

中华人民共和国成立后，在党的“保护、提高、发展”的方针指引下，缂丝产业得到了复兴。1954年，“苏州市文联刺绣生产小组”成立，邀请了两位缂丝资深艺术家沈金水、王茂仙进行缂丝制作。20世纪70年代末，中国实施改革开放，日本商家从中国订购大批量的和服腰带和贵袈衣。缂丝产业迅猛发展，苏州、南通及杭州周边地区缂丝生产厂家和作坊也逐渐成立。20世纪90年代初，由于对外工艺美术贸易日趋下降，缂丝日用品的生产面临企业劳动力过剩、生产萎缩、技艺人员外流，绝大部分企业停业转向，仅苏州刺绣研究所和南通宣和缂丝研制所两家仍继续生产。如今，从事缂丝生产的企业仅有南通宣和缂丝研制所、苏州刺绣研究所有限公司、王金山大师（缂丝）工作室，以及从缂丝制作的老艺术家和新生代民间艺人。图5-6所示为20世纪80 年代王金山缂织的异色异样台屏《寿星图》。1982年，王金山受双面绣的启发，进行了一项大胆的技术革新——双面异缂。首创的是一幅《牡丹·山茶·蝴蝶》，被称为“双面三异”缂丝。三异，即异色、异样、异织，如图5-7所示。

The kossu industry has revived under the guidance of the Party's "protection, improvement and development" policy after the founding of the People's Republic of China. In 1954, the Embroidery Production Group of Suzhou Federation of Literary and Art Circles was established, inviting two veteran artists Shen Jinshui and Wang Maoxian to make Kesi. In the late 1970s, when China's reform and opening up were implemented, Japanese merchants ordered kimono belts and expensive clothes from China in large quantities. Consequently, the kossu industry developed rapidly, and the manufacturers and workshops in Suzhou, Nantong and the surrounding areas of Hangzhou were gradually established. In the early 1990s, due to the declining foreign trade in arts and crafts, the production of kossu daily necessities faced difficulties, such as excessive enterprise labor force, shrinking production, outflow of skilled craftsmen, closing down of most of the enterprises; the production continued only in Suzhou Embroidery Research Institute (SERI) and Nantong Xuanhe Kesi Institute (NXKI).Nowadays, there are only a few veteran artists engaged in making kossu, as well as new generation artists. Still engaged in making kossu are NXKI, SERI Co., Ltd., and Master Wang Jinshan's (Kesi) Studio. Figure 5-6 shows Wang's screen of *the God of Longevity* woven with different colors and patterns in the 1980s. Inspired by double-sided embroidery, in 1982, Wang made a bold technological innovation — double-sided cut silk with different patterns. The first of its kind is a painting of a *Peony, Camellia and Butterfly*, known as double-sided cut silk with three patterns, which are different colors, shapes, and weaves, as shown in Figure 5-7.

图5-6 王金山缂织的异色异样台屏《寿星图》
Wang's Screen of *the God of Longevity*

图5-7 王金山双面异缂《牡丹·山茶·蝴蝶》
Wang's *Peony, Camellia and Butterfly*

缂丝作品一般有三个特点：首先，缂丝作品大多是一种集体创作的作品，后人判断这类作品价值的高低只能看其作品本身的工艺和艺术价值；其次，缂丝的创作往往很费功夫和时间，有时为了完成一件作品需要耗时几个月甚至一年以上，一件缂丝作品的完成往往倾注创作者大量的心血；最后，缂丝作品具有很高的观赏性。缂丝作品一般立体感很强，题材都是人们喜闻乐见的，故其艺术和观赏价值完全可以和名家书画分庭抗礼，甚至有所超越。随着人民生活水平的提高，越来越多的人开始喜爱缂丝织物，促进了缂丝工艺的发展和缂丝织物的生产。

Kossu crafts boast three characteristics. First, they are mostly works of collective creation, judged only by their craftsmanship and artistic value. Second, they are painstaking and time-consuming, and sometimes it takes several months or even more than a year to accomplish one piece. Third, they have great ornamental value. Its artistic and ornamental value can equal and even outweigh the famous artists' calligraphy and painting due to a strong three-dimensional effect, and popular themes. As people's living standards improve, more and more people enjoy kossu products, which promotes its techniques and production.

第三节　宋锦 /Song Brocade

宋朝最著名的丝绸品种是宋锦，是中国传统的丝绸工艺品之一，它的主要产地是苏州，故又称“苏州宋锦”。宋锦色泽华丽，图案精致，质地坚柔，被赋予中国“锦绣之冠”。

The most famous variety of silk in the Song Dynasty was Song brocade, one of the traditional silk crafts in China. Primarily produced in Suzhou, it is also known as Suzhou Song brocade. Gorgeous in colors, patterns, and texture, is known as the most famous brocades in China.

苏州宋锦是在唐朝织锦的基础上发展起来的，蜀锦业更加兴旺发达，果州、保宁府等地所产的生丝源源不断地涌向成都，使用这些地区生产的生丝制作的蜀锦质纹细腻，层次丰富，色泽瑰丽多彩，花纹精致古雅。宋朝，主要是南宋以后，为了满足当时宫廷服饰和书画装帧的需要，织锦得到了极大的发展，并形成了独特的风格，以至于后世谈到锦，必称宋。北宋时，成都转运司在此设立了锦院，专门生产上贡的“八答晕锦”“官诰锦”“臣僚袄子锦”，以及“广西锦”。到南宋时期，成都锦院还生产各种细锦和各种锦被，花色品种更加繁复美丽，这些丝织锦在后来通过贸易等方式逐渐流传到全国，成为著名的传统品种。更重要的是，宋朝织锦吸取了当时成熟的花鸟画中的写生风格，形成具有自己时代独有的，色彩更加复杂的丝织品。明朝宋锦发展到上百余种，苏州织局的“盘条花卉纹绵”以质地柔软，经面整洁而闻名。

Suzhou Song brocade was developed on the basis of brocades in the Tang Dynasty. More developed was Shu brocade produced in Chengdu where the raw silks from Guozhou, and Baoning were assembled to make the brocade, a kind of brocade with attractive texture, layers, colors, and patterns. Brocade in the Song Dynasty was greatly developed, mainly after the Southern Song period, in order to meet the needs of court clothing and decoration of calligraphy and paintings at that time. It formed so unique a style that Song brocade was always referred to when it comes to brocade. In the Northern Song, a brocade workshop was set up in Chengdu by the Department of Transshipment to specialize in the production of such brocades as Badayun, Guangao, Chenliao Aozi, and *Guangxi*. Brocades In the Southern Song, various fine brocades and brocade quilts with more complexity and beauty were also produced in the workshop. These brocade products gradually spread throughout the country by trade and became well–known traditional brocade sources. More importantly, the Song brocade, based on the mature style of sketching flowers and birds at that time. A silk fabric with its unique style and more complicated colors was thus formed. Song brocade in the Ming Dynasty was developed into more than 100 types, with Suzhou Weaving House's brocade with medium–sized and large geometric flower patterns best known for its soft texture and neat warp surface.

大锦是宋锦中最有代表性的品种之一，图案规整，富丽堂皇，质地厚重精致，花色层次丰富，常用于装裱名贵字画。其中重锦最为贵重，特点是在纬线上大量使用捻金线或纯金线，并采用多股丝线合股的长抛梭、短抛梭和局部特抛梭的织造工艺技术，图案更为丰富，常见的图案有植物花卉、龟背纹、盘绦纹、八宝纹等，产品主要是宫廷、殿堂的各类陈设品和巨幅挂轴等。明清之后，宋锦仍然相沿不衰，故宫博物院收藏有一幅清朝重锦《西方极乐世界图轴》，就是宋锦的极品。在2米宽的独幅纹样中，呈现了278个神态各异的人物佛像，还有宫殿巍峨，祥云缭绕，奇花异草，充分展示了重锦高超的艺术技巧。大锦中的细锦在原料选用，纬线重数等方面比重锦简单一些，厚薄适中，易于生产，广泛用于服饰、高档书画及贵重礼品的装饰装帧。

One of the most representative types of Song brocade, Da brocade, is characteristic of neat patterns, dense and exquisite texture, and rich colors, which is often used to decorate famous calligraphy and paintings. Double brocade is valuable in that it is characterized by using a large amount of twisted gold thread or pure gold thread on the weft and adopting the weaving technique of long–throw shuttle, short–throw shuttle, and special shuttle processing. Its patterns are more abundant, such as plants and flowers, tortoise back, geometric, and treasures. Its products are mainly various furnishings and giant hanging shafts in royal palaces and halls. After the Ming and Qing Dynasties, the development of Song brocade continued, a period of history witnessing a great Qing Double brocade painting scroll *the Sukhavati Brocade Scroll*, a priceless masterpiece of Song brocade preserved at the Palace Museum.Presented in the scroll are 278 statues of Buddha with different expressions in a 2-meter-wide single-frame pattern, as well as the grand palace, bright clouds, strange flowers, and plant, demonstrating the superb artistic skills of *Double brocade*. Unlike making *Double brocade*, Fine brocade is less demanding in terms of raw material selection and weft weight, with moderate thickness, easy production, and wide application to garments, expensive calligraphy, paintings, and gifts.

宋锦图案具有浓烈的民族风格，表现了中国人民对美好生活的向往，表达了民族意识和民族心理，含蓄而文雅。宋锦的图案一般以几何纹为骨架，内填以花卉、瑞草，或八宝、八仙、八吉祥。八宝指古钱、书、画、琴、棋等，八仙是扇子、宝剑、葫芦、柏枝、笛子、绿枝、荷花等，八吉祥则指宝壶、花伞、法轮、百洁、莲花、双鱼、海螺等。

The Song brocade pattern has a strong national style, showing the Chinese people's expectation for a better life, expressing the national consciousness and mentality as subtle and elegant. The patterns are generally based on geometric, filled with flowers, grass, eight treasures, eight immortals' objects, and eight auspicious symbols. The eight treasures refer to ancient coins, books, paintings, musical instruments and chess, etc., eight immortals' objects fan, sword, gourd, cypress twig, flute, green twig, lotus, etc., and eight auspicious symbols to treasure pot, umbrella, dharma wheel, auspicious knot, water lily, double fish, conch, etc.

宋锦的配色独具匠心，别有风味。它的色彩特点是高雅清淡，古朴含蓄。仿古宋锦在进行配色时，一般都含灰色调，它要求各色的色纯度、色明度是统一而和谐的，因而，配制的色彩柔和平稳，多用调和色，一般很少用对比色，给人一种古旧而儒雅的感觉。宋锦的地纹色，大部运用米黄、蓝灰、泥金、湖色等。花纹图案则分三种类型，互相搭配；较大的花纹用庄严而稳重的常用色调；主花的花蕊或图案用比较温和而鲜艳的特用色彩；而配合花朵的包边或分隔上两类色彩的小花纹则用协调而中和的间色。三类色彩经过巧妙的配合，形成宋锦庄严美观的效果：灰而不闷、丽而不刺、繁而不乱、活泼自然、古色古香，如图5-8所示。

Song brocade is unique and ingenious in terms of color scheme, namely elegance, simplicity, and quaintness. It usually contains gray tones when color scheming is processed, which requires purity and brightness of each color, thus ensuring that the color design is unified and harmonious. Therefore, the prepared colors are soft and smooth, with more use of compound color and less use of contrast, showing a sense of antiquity and elegance. The color of its ground weave mostly uses rice yellow, blue ash, mud gold, the lake green. Divided into three types, the patterns are well matched: larger patterns are made of solemn common tones, the stamens or patterns of the main flowers of relatively mild and bright ones, small patterns matching with the edges of the flowers or dividing the above two types of colors of coordinated and neutral ones. The three types of colors are skillfully coordinated to form a solemn and aesthetic effect of the brocade: gray but not boring, pleasant but not abrupt, complicated but not chaotic; lively, natural, and quaint, as shown in Figure 5-8.

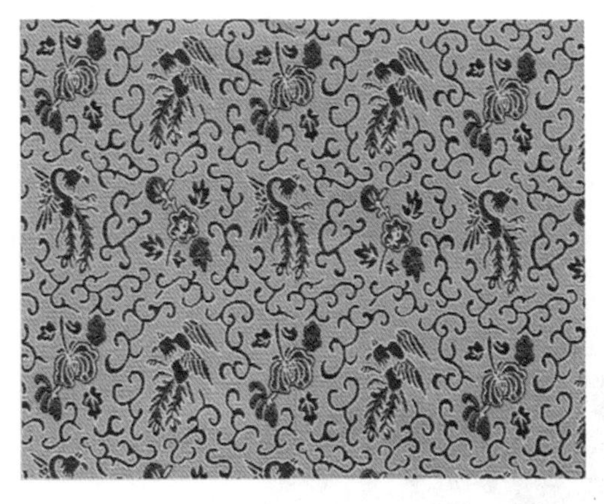

图5-8　宋锦
Song Brocade

宋锦不仅具有不菲的收藏价值，还解决了其他丝绸类手工艺品无法实现的实用性问题。宋锦的工艺决定了它的实用性，质地柔软坚固、图案精美绝伦、耐磨且可以反复洗涤，适用面非常广泛。结合了传统的制作工艺和现代的审美观念，将宋锦创新地应用到箱包、服装、家纺、工艺品等众多领域，使原本只能用作书画装裱的宋锦真正走入了寻常百姓家。图5-9所示为常见的宋锦产品。

Soft and sturdy, Song brocade is skillfully patterned, wear-resistant, and wash-resistant. In addition to its collection value, it is superior to any other silk craft because of its exquisite craftsmanship and wide range of applications. Coupled with traditional techniques and modern aesthetic concepts, it has been innovatively applied to many fields such as bags, clothing, home textiles, handicrafts, and so on. Once only used as a painting and calligraphy decoration, it has now been made available as life necessities. As shown in Figure 5-9, this is a piece of commonly seen Song brocade products.

图5-9　常用宋锦产品
Commonly Used Song Brocade Products

2006年，宋锦被列入第一批国家级非物质文化遗产名录，传承单位为苏州传统丝绸非遗馆（图5–10）。2009年，宋锦织造技艺作为中国传统桑蚕丝织技艺入选联合国教科文组织《人类非物质文化遗产代表作名录》。

Song brocade, recommended by Suzhou Traditional Silk Intangible Cultural Heritage Museum, was inscribed in 2006 on the representative list of Chinese intangible cultural heritage, as shown in Figure 5–10. The song brocade weaving technique as the Chinese Traditional Silk Weaving Technique was inscribed in 2009 by UNESCO on the *Representative List of the Intangible Cultural Heritage of Humanity*.

图5–10　苏州传统丝绸非遗馆
Suzhou Traditional Silk Intangible Cultural Heritage Museum

2014年11月，在北京雁栖湖国际会议中心举行的亚太经合组织（APEC）晚宴上，与会代表身着设计新颖的中国特色服装抵达会场，他们穿的宋锦"新中装"面料便是产自苏州吴江鼎盛丝绸有限公司生产，如图5–11所示。

At the APEC banquet held at the international conference center of Yanqi Lake in Beijing in November 2014, all the participants arrived at the venue in Chinese–style outfits with innovative designs and took a great group photo.The Song brocade used to make the New Chinese Tunics was produced by Wujiang Dingsheng Silk Corporation Limited in Suzhou, as shown in Figure 5–11.

图5–11　宋锦"新中装"
Song Brocade in the New Chinese Tunics

第四节　旗袍 /Cheongsam

旗袍，中国和海外华人女性的传统服装，被誉为中国国粹和女性国服，是中国悠久的服饰文化中最绚烂的现象和形式之一。旗袍属于袍服中的一种，它是由满族旗人的袍服演变而来。旗人是对满族八旗人的通称，满族女性的长袍就成了人们常说的“旗袍”。女性旗袍在清朝极为盛行，在当时的中原和南方地区的其他民族的女性，都渐渐地穿上了类似的长袍马褂，上袄下裙的衣服。旗装大多采用平直的线条，衣身宽松，下摆不开衩，胸腰围度与衣裙的尺寸比例较为接近；在袖口领口有大量盘滚装饰。旗装色彩鲜艳复杂，用料等花色品种多样，对比度高的色彩搭配更容易被人接受。在领口、袖头和掖襟上添加几道鲜艳花边或彩色牙子盘滚设计。旗装是一种平面服饰，盘滚成为旗装除面料外的唯一设计空间，因而以多盘滚为美。

A traditional costume of Chinese and overseas Chinese women, cheongsam, also known as Chinese dress or Qipao, is one of the most brilliant phenomena and forms of Chinese garment culture. Cheongsam belongs to one of the robes, which evolved from the cloth of the Manchu ethnic minority. People from the Manchu Eight Banners are commonly known as bannermen and the robes of Manchu women are thus called cheongsam. Cheongsam gained great popularity in the Qing Dynasty, with the women of other ethnic groups in the regions of central plains and the south wearing clothes similar to the robes. Cheongsam mostly adopts straight lines, a regular fit, with no slit on both sides. The size of its chest and waist is proportionally close to that of the lower part, with intensified traditional Chinese decorative techniques on the cuff and necklines such as knotted or frog button. Bright and complicated, the colors of the robes are diverse, with high-contrast colors more accepted. Bright-colored lace or embedded fabrics are added to the collar, sleeves, and tucks with knotted or frog button and contrast trim, two of the major techniques to make a piece of cheongsam. A single-piece ornately cut dress, cheongsam boasts its knotted or frog button decoration.

19世纪40年代，上海华洋杂居，西方面料、缝纫方式及外国服装款式开始进入国内，传统的民间服饰渐渐有了变化。起先，上海妇女将旧式长裙加以改造，成为长马甲（称为旗袍马甲），上身仍是短袄。20世纪30年代前后，受欧美短裙影响，原来长短适中的旗袍开始变短，下摆上缩至膝盖，袖口变短变小。后来又出现了校服式旗袍，下摆缩至膝盖以上1寸，袖子采用西式。20世纪30年代初，旗袍又开始变长，下摆下垂。20世纪30年代中期，发展到极点，袍底落地遮住双脚，称为“扫地旗袍”。原来能遮住手腕的旗袍袖子缩短至肘部。以后袖长越来越短，缩至肩下两寸，如今几乎无袖。

Shanghai’s founding as a trading port in the 1840s, coupled with mixed communities, led to changes in traditional Chinese folk costumes, since Western materials, sewing methods, and foreign clothing styles began to enter China. At first, women in Shanghai transformed the old-style dress into a long waistcoat (called a Cheongsam waistcoat), with the upper body remaining short. In the 1930s, under the influence of the European and American skirts, the medium-length Cheongsam began to shorten, with the sweep shrinking up to the knee and the cuffs shortening. Later, there was a school uniform Cheongsam, its sweep shrinking to 1 inch above the knee, and the sleeves keeping Western style. After the 1930s, it began to lengthen, with sweep drooping. In the mid-1930s, the sweep covered the feet, known as floor sweeping cheongsam. The sleeves of Cheongsam that originally covered the wrist were shortened to the elbow. They became shorter and shorter, shrinking to two inches under the shoulders, and almost sleeveless today.

如图5-12所示，旗袍通过缩短长度、收紧腰身、提高开衩的改良，逐渐形成中西合璧的服装款式，以一种含蓄的形式表现性感，从而迎合了现代时尚。旗袍非常适合表现中国妇女的体形和贤淑的个性、东方民族的优雅气质，所以一直受到国内外女性的青睐和赞赏，成为中国最具代表性的服饰。

As shown in Figure 5-12, Cheongsam gradually formed a mixture of Chinese and Western clothing styles by shortening the length, tightening the waist, and slitting higher, which shows the sexiness in an implicit way, thus catering to modern fashion. Suitable for showing the body shape and virtuous personality of Chinese women and the elegance of the oriental nation, it has been favored by women at home and abroad, becoming the most representative clothing in China.

图5-12　旗袍
Cheongsam

常见的大红色旗袍，色彩绚丽醒目，款式别致，充分展现出中华民族悠久的历史文化，着重体现东方女性含蓄优雅的魅力。现代常见的旗袍织锦缎，图案为传统的中国纹饰如双鱼、富贵花、梅花等，还有以中国水墨画手法描绘的花卉图案来设计的手绘旗袍。旗袍样式也十分丰富，如图5-13所示。

The common red Cheongsam, with bright colors and distinctive styles, fully demonstrates the long history and culture of the Chinese nation, highlighting the subtle and elegant charm of oriental women. Modern Cheongsam tapestry satin boasts traditional Chinese patterns such as double fishes, peony, and plum blossom. There is also hand-painted Cheongsam with Chinese ink painting design. Its styles are also abundant, as shown in Figure 5-13.

图5-13 旗袍的样式
Style of Cheongsam

旗袍也逐渐走向了世界，无论是在国际时装舞台，还是日常工作和生活，旗袍以多变的姿态展现着女性美，演绎着别样的东方风情。

cheongsam is made globally available. Whether it is on the international fashion stage or in daily work and life, it shows women's beauty in a variable posture, demonstrating a different oriental style.

参考文献/References

［1］钱小萍. 中国织锦大全：民族织锦编［M］. 北京：中国纺织出版社，2014.

［2］赵丰. 中国丝绸通史：The general history of Chinese silk［M］. 苏州：苏州大学出版社，2005.

［3］唐林. 蜀锦与丝绸之路［J］. 中华文化论坛，2017，6（3）：20-25.

［4］黄能馥，陈娟娟. 中国丝绸科技艺术七千年：历代织绣珍品研究［M］. 北京：中国纺织出版社，2002.

［5］吴方浪. 汉代“蜀锦”兴起的若干原因考察［J］. 丝绸，2015，52（9）：72-76.

［6］赵丰. 中国丝绸艺术史［M］. 北京：文物出版社，2005.

［7］刘治娟. 丝绸的历史［M］. 北京：新世界出版社，2006.

［8］刘行光. 丝绸［M］. 重庆：西南师范大学出版社，2014.

［9］邢声远. 丝绸的故事：技术与文化［M］. 济南：山东科学技术出版社，2020.

［10］路甬祥，钱小萍. 中国传统工艺全集：丝绸织染［M］. 郑州：大象出版社，2005.

［11］黄君霆，朱万民，夏建国. 中国蚕丝大全［M］. 成都：四川科学技术出版社，1996.

［12］李晓，李俊久. “一带一路”与中国地缘政治经济战略的重构［J］. 世界经济与政治，2015（10）：30-59，156.

［13］张保丰. 中国丝绸史稿［M］. 上海：学林出版社，1989.

［14］郭鹏，冯苓. 丝绸面料织造工艺及产品的概述［J］. 江苏丝绸，2021，50（2）：12-20.

［15］张剑锋. 试论明清官营丝绸织造及其织物典制［J］. 纺织报告，2019（5）：53-60，64.

［16］《丝绸文化与产品》编写组. 丝绸面料的染整（1）：丝绸印染前处理［J］. 现代丝绸科学与技术，2018，33（1）：28-29.

［17］陈文兴. 触蒸前处理提高生丝洁净的探讨［J］. 纺织学报，1988，9（5）：17-20.

［18］刘慧，徐英莲. 鲜茧前处理工艺对缫丝及生丝性能的影响［J］. 现代纺织技术，2018，26（1）：17-21.

［19］宁肇棠. 国内外真丝绸染整技术现状简介［J］. 印染助剂，1988，5（3）：3-5.

[20] 孔昱萤，关晋平，吴家宝，等. 酸性染料染色真丝绸的剥色技术探究 [J]. 现代丝绸科学与技术，2019，34 (5)：11-15，21.

[21] 董玮，庄德华，何瑾馨. 叔胺类化合物与K型活性染料的反应性能研究及其在真丝染色上的应用 [J]. 印染助剂，2007，24 (7)：25-28.

[22] 黄雪红. 真丝条纹绸的染整工艺设计 [J]. 现代丝绸科学与技术，2012，27 (3)：98-101.

[23]《丝绸文化与产品》编写组. 丝绸面料的染整 (5)：丝绸整理技术 [J]. 现代丝绸科学与技术，2018，33 (4)：37-38.

[24] 李祥，高光东，杜韩静，等. 真丝织物洗可穿整理加工 [J]. 染整技术，2007，29 (10)：22-23，46，58.

[25] IanHolme，徐锡环，马建兴. 真丝绸染整技术 [J]. 国外丝绸，2006，21 (3)：15-16.

[26] 周宏湘. 真丝绸染整新技术 [M]. 北京：纺织工业出版社，1997.

[27] 徐昳荃，张克勤，赵荟菁. 文化丝绸 [M]. 苏州：苏州大学出版社，2016.

[28] 罗永平. 江苏丝绸史 [M]. 南京：南京大学出版社，2015.

[29] 袁宣萍，赵丰. 中国丝绸文化史 [M]. 济南：山东美术出版社，2009.